Dean Rood
A Practical Guide to the Care, Maintenance, and Troubleshooting
of Capillary Gas Chromatographic Systems

Chromatographic Methods

Series editors

W. Bertsch
H. Frank
W. G. Jennings
P. Sandra

A Practical Guide to the Care, Maintenance, and Troubleshooting of Capillary Gas Chromatographic Systems

By Dean Rood

2nd enlarged and revised Edition

Hüthig Verlag Heidelberg

Dean Rood
J & W Scientific
Specializing in High Resolution Chromatography
91 Blue Ravine Road,
Folsom, CA 95630-4714
(916) 985-7888, USA

Die Deutsche Bibliothek – CIP-Einheitsaufnahme

Rood, Dean:
A practical guide to the care, maintenance, and troubleshooting
of capillary gas chromatographic systems / by Dean Rood. –
2., enlarged and rev. ed. – Heidelberg: Hüthig. 1995
 (Chromatographic methods)
 ISBN 3-7785-2319-8

© 1995 Hüthig GmbH, Heidelberg
Printed in Germany
ISBN 3-7785-2319-8

PREFACE

Far from suffering the stagnation that some have predicted, the field of gas chromatography continues the evolutionary process. This is well demonstrated by the continuous series of developments - in columns, equipment, apparatus, techniques and applications - that have occurred since the publication of the first edition of this very successful offering. Problems experienced by uses are different in some cases, and these differences sometimes necessitates different approaches to care, maintenance and troubleshooting.

The Troubleshooting Group at J&W Scientific receives an average of 1,000 telephone and telefax queries each month; I am highly pleased that one of our senior chemists has taken the time and effort to share this information and advice based on this vast experience. I fully expect that this second edition will prove to be an even more valuable aid to the practicing analyst. I am very pleased to add my endorsement to this effort.

Walter Jennings
Professor Emeritus

ACKNOWLEDGEMENTS

A new edition of a book is often more than just changes and additions due to the advent and development of new technologies and techniques since the birth of the preceding manuscript. In my case, a bit of hindsight and a few more years of experience played a part in influencing the changes, additions and improvements in this 2nd edition. More importantly, discussions with my colleagues and especially other chromatographers around the world have significantly influenced the content and form of this book. Your comments, suggestions, ideas, experiences and problems (along with mine) over the years have supplied me with the bulk of the material. Thank you for the help and inspiration.

I would like to thank Michael Feeney, Kelly Turner, Daron Decker, Marc Dinnauer and Julie Knitter for their suggestions, recommendations and careful reading of the manuscript for the first edition. A large part of that edition still resides within these pages. Many thanks to Tony Rackstraw of Hüthig for his patience while I continued to promise the delivery of this manuscript and the corrections to my continued butchering of the English language. All of the errors, mistakes and goofs belong to me.

Most of all, all my love and gratitude to my wife, Pam, and my son, Christopher. They endured and survived (again) many weekends and evenings of my inattention, anguish and occasional cursing. Again, my sincere thanks to everyone - without your help and support this book would be just another unfulfilled intention.

TABLE OF CONTENTS

2 Basic but Essential Definitions and Equations

3 Capillary GC Columns: Tubing

4 Capillary GC Columns: Stationary Phases

5 Capillary GC Columns: Dimensions

6 Carrier Gases

7 Operation and Maintenance of Injectors

8 Operation and Maintenance of Detectors

9 Column Installation

10 Column Test Mixtures

11 Causes and Prevention of Column Damage

12 Troubleshooting Capillary GC Systems

Appendix

INTENTIONS AND INTRODUCTION

Intentions and Introduction

There already seems to be a number of excellent references on gas chromatography (GC), so why this book? Well, there are several reasons. There is a large number of gas chromatographs in use. If is often stated that gas chromatography is the most common instrumental analytical technique in routine use. The availability of easy to operate, affordable and feature laden instruments has made GC a powerful analytical technique accessible to nearly every laboratory.

Commercially available capillary columns of high quality have existed for less than 20 years. For a number of reasons, many GC users are not extremely experienced in the practice of capillary gas chromatography. Many of these users do not possess a level of comprehension of the technique that allows them to prevent and solve many of the problems that commonly occur. Much of this comprehension comes from years of experience and the problems that accompany that experience. The combination of accessible instruments and capillary columns along with inexperienced users has created the need for practical information on the care, maintenance and troubleshooting of capillary columns and instruments.

One of the goals of this book is to provide practical information that will maximize both capillary column lifetime and the performance of the gas chromatographic system. The other goal is to provide an efficient and logical troubleshooting guide with the real intention to reduce or prevent performance breakdown problems from occurring. An in-depth knowledge of chemistry and chromatography (and other foreign languages) is not required. This book, in no shape or form, attempts to thoroughly explain every detail about capillary gas chromatography; it is intended as a practical guide so that the urge to hit the GC with a hammer as a last resort does not occur. In-depth technical information about GC techniques, instrumentation, specific applications and other gory details can be found in the books listed in the reference section.

Many generalizations and simplifications have been exercised to keep the information in a basic and widely digestible form. Again, this book is intended for the average GC user and not those whose entire life revolves around capillary gas chromatography. The topics covered within these pages are based on the most common problems, questions and misconceptions about capillary gas chromatography. These topics have been assembled and presented in an unique, practical and concise format suitable even for the most inexperienced GC user.

1 INTRODUCTION TO CAPILLARY GAS CHROMATOGRAPHY

1.1 A Brief and Simple Introduction to Gas Chromatography

1.1.1 What is Gas Chromatography?

In a broad sense, gas chromatography is a very powerful and one of the most common instrumental analysis techniques in use. When properly utilized, it provides both qualitative (i.e., what is it?) and quantitative (i.e., how much?) information about individual components in a sample. Gas chromatography involves separating the different compounds in a sample from each other. This allows the easy identification and measurement of the individual compounds in a sample. The compounds are separated primarily by the differences in their volatilities and structures. Many compounds and samples are not suitable for gas chromatographic analysis due to their physical and chemical properties.

1.1.2 What Types of Compounds are Suitable for GC Analysis?

For a compound to be suitable for GC analysis, it must possess appreciable volatility at temperatures below 350-400°C. In other words, all or a portion of the compound molecules have to be in the gaseous or vapor state below 350-400°C. Another characteristic is the compound must be able to withstand high temperatures and be rapidly transformed into a vapor without degradation or reacting with other compounds. Unfortunately, this type of information about a compound is not readily available in references or other sources; however, some estimates and generalizations can be made from the structure of the compounds.

Compound structure and molecular weight can be used as indicators of potential GC analysis suitability. The actual characteristic of interest is compound volatility. Compounds with very low volatilities are not suited for GC analysis since they do not readily vaporize. Compound boiling points are not always good indicators of volatility. There are many high boiling compounds that can be analyzed by GC. As a general rule, the greater the molecular weight or polarity of a compound, the lower its volatility. Both factors have to be considered. For example, a large, non-polar compound may be more volatile than a small, polar compound. Also, one polar group on a large molecule has less of an influence than one polar group on a small molecule.

Hydrocarbons with molecular weights over 500 are routinely analyzed using standard GC systems, and hydrocarbons with molecular weights over 1400 have been easily analyzed using the properly equipped GC and type of column. The presence of polar functionalities such as hydroxyl and amine groups severely decrease compound volatility. Some small molecules such as sugars and amino acids can not be easily analyzed by GC due to the large number of polar groups.

As a rule, inorganic compounds are not suitable for GC analysis. Metals and salts do not possess the required volatility. Many organometallics have sufficient volatility for analysis due to the high organic content of these molecules. Most organic compounds are suitable for GC analysis; however, there are many exceptions. Many biomolecules and pharmaceuticals are thermally sensitive and degrade at the temperatures used in gas chromatography. Some compounds react with the materials used in gas chromatographs and columns and can not successfully analyzed by GC. There are no realistic, absolute guidelines that can be used to determine whether a compound can be analyzed by GC. Overall, it has been estimated that only about 10% of all compounds can be analyzed by GC.

1.1.3 The Basic Parts of a Gas Chromatograph

A gas chromatographic system is comprised of six major components: gas supply and flow controllers, injector, detector, oven, column, and a recording device (Figure 1-1). In most cases, the injector, detector and oven are integral parts of the gas chromatograph; the column, gases and recording device are separate items and are often supplied by a different manufacturer. All of the components are further described in individual sections or chapters with the exception of the oven and recording devices.

Figure 1-1: Simple Diagram of a Typical Gas Chromatograph

1.1.3.1 Gas Supply and Flow Controllers

High purity gases are supplied from a pressurized cylinder or gas gener-
ator. Pressure regulators on the cylinders or generators control the amount
of gas delivered to the gas chromatograph. Flow controllers or pressure
regulators in the gas chromatograph control the flow of the various gases
once they enter the instrument.

The column is installed between the injector and detector. Gas at a
precisely controlled flow is supplied to the injector; this gas is called the
carrier gas. The carrier gas flows through the injector and into the open
tubular column. The gas travels the length of the column and exits through
a detector. To function as desired, most detectors require specific gases at
the proper flow rates.

1.1.3.2 Injector

The injector introduces the sample into the open tubular column. The
injector is a hollow, metal cylinder containing a glass liner or insert. The
column is inserted into the bottom of the injector so that the column end
resides inside of the glass liner. Sample dissolved in a solvent (sometimes

pure material or a gas) is introduced into the injector through a resealable septum using a small syringe. The injector is heated to 100-300°C, thus any volatile sample components are rapidly transformed into a vapor. The carrier gas mixes with the vaporized portion of the sample and carries the sample vapors into the column.

1.1.3.3   Capillary Column and Oven

The column resides in an oven whose temperature is accurately controlled. If unimpeded, vaporized compounds move through the column at the same rate as the flowing carrier gas. However, the interior walls of columns are coated with a thin film of polymeric material called the stationary phase. This stationary phase impedes the movement of each compound down the column by a different amount - this behavior is called retention.

The length and diameter of the column, the chemical structure and amount of the stationary phase, and the column temperature all affect compound retention. If all of these factors are properly selected, each compound travels through the column at a different rate. This makes the compounds exit the column at different times. As each compound leaves the column, its presence and amount are measured by the detector.

1.1.3.4   Detector

As each compound exits the column, it enters the detector. The detector interacts with the compounds based on some physical or chemical property. Some detectors respond to every compound while others respond only to a select group of compounds. The interaction generates an electrical signal whose size corresponds to the amount of the compound. The detector signal is then sent to a recording device for plotting.

1.1.3.5   Recording Device

Common recording device includes strip chart recorders, integrators and computer data systems. They vary in their degree of complexity, features and price. The recording device plots the size of the detector signal versus the time elapsed since sample introduction into the injector. The plot is called a chromatogram and appears as a series of peaks (Figure 1-2).

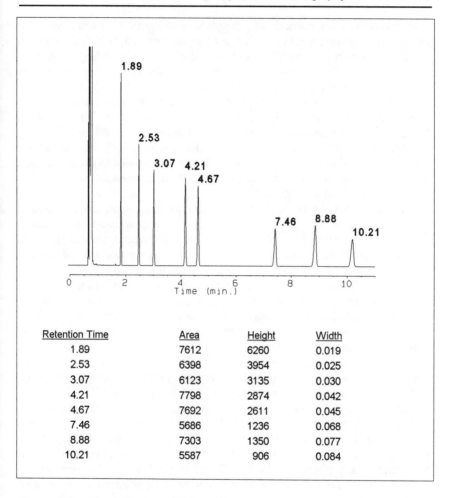

Figure 1-2: Chromatogram and Report

1.1.4 The Chromatogram

In the ideal situation, each peak in the chromatogram represents a single compound in the sample. It is not unusual for more than one compound in a sample to interact with the column in the same manner, thus each compound has the same retention. This results in a single peak that represents more than one compound (complete co-elution). In some cases, the interactions are very similar. This results in two peaks that partially overlap (partial co-elution). Using the proper column and operating conditions minimizes dual peak identities or overlapping problems, but there are cases where complete separation is not possible.

Each peak in the chromatogram is assigned a retention time. It is the time required for a compound to travel through the column. The recording device usually calculates and prints the retention times and size for each peak on the chromatogram or in a table (Figure 1-2). Retention times are usually reported in minutes and the peak size in an unitless area or height value. Strip chart recorders require the manual measurement of the retention times and peak sizes.

If a peak represents only one compound, identifying the compound corresponding to the peak is easy. By injecting a prepared solution containing the compound (commonly called a standard), the compound's retention time is determined. Using the same column and conditions, the sample is analyzed. If a peak in the sample has the same retention time as the one in the standard, there is a good probability that the sample contains the compound. If a peak in the sample does not have the same retention time, the sample does not contain the compound. There are some cautions that have to be exercised to prevent possible misidentification of peaks, and these are discussed in Section 1.2.

To determine the amount of a compound in the sample, the size of its peak in the chromatogram is used. The size of a peak is directly proportional to its amount in the sample or standard. By injecting a standard containing a known amount of the compound of interest, a peak of known size and retention time is obtained. The size of the peak in the sample is compared to the size of the corresponding peak in the standard. A simple ratio is set up for quantitation. For example, if the peak in the sample is two times larger than the peak in the standard, the injected portion of the sample contains two times the amount of the compound than the amount known to be present in the standard.

1.2   Limitations and Cautions

Just like any other analytical technique, gas chromatography has some limitations and shortcomings. A peak in the sample chromatogram with the same retention time as a peak in a standard chromatogram does not abso-lutely confirm the identity of the corresponding compound. There are many compounds that have the same retention times for the same column and conditions. Many times these other compounds are unknown components of the sample, vary in amount between individual samples, and are not the compounds being measured as part of the analysis. The presence of these other compounds are often outside of the control of the analyst and can not be determined before the GC analysis. These reasons often make positive identifications less than absolute if only the retention times obtained with one column are used as the basis of compound identification.

Samples containing many peaks other than those of interest, chromatograms with potentially ambiguous peak identifications, and samples of largely unknown composition often require confirmatory information. In some cases, re-analyzing the standard and sample using a column with a different stationary phase confirms the compound identities. The same result with two different stationary phases provides a high level of confidence in the peak identifications. Other situations may require specialized detectors such as mass spectrometers and infra-red spectrophotometers. These types of detectors provide structural information in additional to the chromatographic data. This type of analysis is considered to be among the strongest of the compound identity confirmation techniques.

Many samples or compounds can not be directly analyzed by GC, but can be converted or treated to make them suitable for analysis. This may involve chemically converting the compounds into a more suitable form for analysis (i.e., derivatization), or removing unwanted or harmful sample components (i.e., extraction). Derivatization is used to improve the volatility of some compounds or to increase the size of the peaks. Extractions are often used to remove a compound from a complex matrix (e.g., blood, soil, sludge, food products) which reduces the number of unwanted peaks in a chromatogram. Also, extractions are often used to concentrate the sample making it easier to detect the compounds. Failure to properly treat samples prior to analysis often results in incorrect data, damage to the GC system or the inability to correctly perform the GC analysis.

1.3   The Mechanism of Compound Separation

How does the column work? What happens inside the column? How do the compounds move through the column? Why do some compounds stay in the column longer than others? How does the sample get into the column? These are some of the most basic questions asked about gas chromatography. Knowing the answers does not automatically make a chromatographer produce better results, but the knowledge is very valuable in solving and preventing problems, selecting columns, and understanding unexpected results. Complicated discussions involving thermodynamics and molecular interactions are necessary to fully answer these questions. Fortunately, comprehension at this level is not necessary to become an excellent chromatographer. A basic understanding of the concepts, and not the intricate details, provides a chromatographer with all of the information necessary to produce the most consistent, trouble free and best results.

1.3.1 A Very Simple Description of the Chromatographic Process

The separation of a sample into its individual compounds by a capillary GC column can be described by a very simple concept. The sample containing a mixture compounds enters the column and collects in the front of the column (Figure 1-3a). Then the molecules of each compound start to move down the column at a different rate (Figure 1-3b). The fastest moving molecules reach the end of the column first, enters the detector, thus corresponding with the first peak in the chromatogram (Figure 1-3c). The

Figure 1-3:   Separation of the Sample in the Column

next fastest compound molecules follows, and this process continues until all of the remaining compounds have left the column (Figure 1-3d). Since the compounds each leave the column at different times, they are separated. Any compounds that travel through the column at the same rate are not separated and have the same retention times.

1.3.2 A Detailed Description of the Chromatographic Process

Capillary columns are composed of three distinct parts. The tubing is fused silica (glass) with an outer protective coating. The inner walls are coated with a thin film of polymeric material called the stationary phase. The sample compounds interact with the stationary phase, and this interaction is responsible for the separation properties of the column.

Once in the column, the molecules for each compound distribute between the mobile phase (carrier gas) and the stationary phase (Figure 1-4a). Molecules in the mobile phase move down the column; molecules in the stationary phase do not move down the column (Figure 1-4b). The carrier gas transports the compound molecules down the column. Simultaneously, the molecules are moving in a random, latitudinal motion. Eventually, each molecule comes into contact with the stationary phase. Each one enters the stationary phase when this occurs. For every molecule entering the stationary phase, another one leaves the stationary phase to take its place in the mobile phase (Figure 1-4c). This maintains the same overall distribution of the molecules between the two phases.
The process of exchange between the phases is repeated thousands of times for each molecule. The net effect is the movement of the molecules down the column (Figure 1-4d).

The rate of molecule movement down the column depends on the distribution of the molecules between the stationary and mobile phases. The greater the percentage of molecules in the mobile phase, the faster the molecules travel down the column. This results in a short residence time for the molecules in the column and a short retention time for the corresponding peak. Separation of two compounds occurs when the distribution of their molecules between the stationary and mobile phases are different. If the distributions are the same, co-elution occurs.

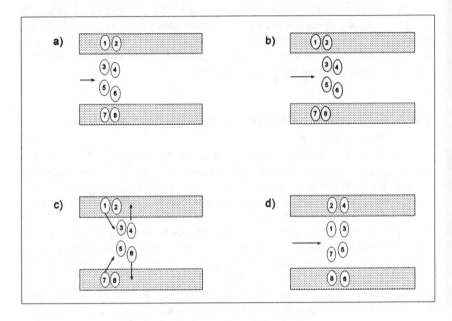

Figure 1-4:   Movement of Molecules Down the Column

The distance or time between the various groups of molecules (with each group representing one compound) as they exit the column determines the amount of separation between the peaks. While this separation distance is important, there is more to chromatography than just separation. The length of column occupied by the molecules for each compound is critical. A narrow band of compound molecules occupying a short length of column is desired. If the width of the molecule bands is narrow, a large separation between the band of molecules is not needed to prevent overlap of the different compound molecules (Figure 1-5a). If the width of the molecule bands is broad, the same amount of separation results in an overlap of the different compound molecules (Figure 1-5b). When the molecule bands are broad, greater separation is need to prevent overlapping of the molecule bands (Figure 1-5c).

Figure 1-5: Width of the Molecule Bands and the Effect on the Chromatogram

1.4 Factors Affecting Separation

The distribution of compound molecules depends on the stationary phase, the compound and the column temperature. Only these three factors influence the amount of peak separation (i.e., distance or time between the peaks). Other factors such as column dimensions and carrier gas do not have a direct affect on separation.

1.4.1 Stationary Phase

It can be viewed that compound molecules can be regarded as dissolving in the stationary phase. For a particular compound, its molecules may be more soluble in one stationary phase than in another. In this case, the compound's molecules distribute differently in each stationary phase (Figure 1-6). This means different retention is obtained with each stationary phase. For multiple compounds, the change in retention (i.e., molecule distribution) for each compound is usually different. For this reason, co-eluting compound peaks for one stationary phase often separate with a different stationary phase.

Figure 1-6:   Molecule Distribution in the Mobile and Stationary Phases - A Change in the Stationary Phase

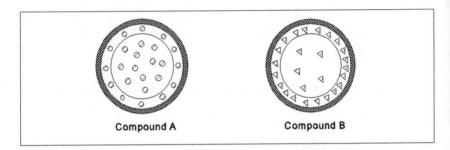

Figure 1-7:   Molecule Distribution in the Mobile and Stationary Phases - A Change in the Compound

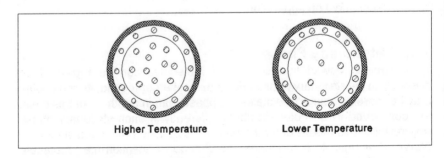

Figure 1-8:   Molecule Distribution in the Mobile and Stationary Phases - A Change in the Column Temperature

1.4.2 Compound Structure

The molecules for different compounds distribute in a different ratio in the same stationary phase (Figure 1-7). Different compounds have different solubilities in the same stationary phase. If this occurs, peak separation is obtained. If two compounds have the same distribution in the stationary phase, the corresponding peaks do not separate. This is the reason for co-eluting peaks on the same column.

1.4.3 Column Temperature

The distribution of molecules between the stationary and mobile phases depends on column temperature. At higher column temperatures, there are fewer molecules in the stationary phase and more in the mobile phase than at a lower temperature (Figure 1-8). The presence of fewer molecules in the stationary phase results in faster migration down the column and shorter retention times. This accounts for longer retention times at lower column temperatures and shorter retention times at higher column temperatures.

At higher column temperatures, the differences in the distribution of different compound molecules become smaller. Smaller distribution (retention) differences in the various compound molecules result in less separation between the corresponding peaks. In general, this results in a decrease in separation at higher column temperatures (less difference in the molecule distributions) and a separation increase at lower column temperatures (greater differences in the molecule distributions).

1.5 A Few Final Comments

It is important to realize that the quality of a chromatographic separation is limited by the performance of the weakest component of the chromatographic system whether it be the column, operational parameters or the instrument itself. There are numerous factors influencing the separation of two compounds.

A basic understanding of how all of the components affect the results is very important; troubleshooting and system operation become much easier and more efficient. Simple and efficient elimination of the possible problem sources through an understanding of the interaction of the entire GC system substantially increases the ease with which troubleshooting can be successfully completed. More importantly, procedures to prevent or minimize the frequency and severity of problems can be implemented so that troubleshooting becomes an infrequent occurrence.

Column and instrument performance and quality have improved significantly over the past 10 years. Rarely are they the actual source of performance problems. The demands on the GC analyst continues to increase as the information provided by GC systems becomes more detailed, precise and accurate. Problems and variations become more obvious, measurable and significant. The demand continues to grow for the analysis of more compounds, at lower levels and in more complex and different matrices. With all of this comes the need to better comprehend, control and optimize the operation of capillary GC systems.

2   BASIC DEFINITIONS AND EQUATIONS

2.1 Why Bother?

There are a number of terms that are commonly used to describe various chromatographic and column characteristics, behaviors, and conditions. An understanding of these terms on a basic level is helpful for comparing column performance, determining column quality, troubleshooting, interpreting the test mixture chromatograms, and comprehending the fundamentals of chromatographic theory.

2.2 Peak Shapes
2.2.1 Peak Width (W)

Depending on the compounds, conditions and column, the peaks in a chromatogram have different widths. Peak widths are important indicators and measures of column and system performance. There are two peak width measurements that are commonly used - the width at the base and at half height.

Peak width at the base is measured by extending the peak sides down to the baseline and measuring the width at the baseline (Figure 2-1a). This can be uncertain if the baseline is not level or if the peak is partially co-eluting with another one. Measuring the peak width at half height is usually easier. This involves measuring across the peak half way up its height instead of at the base (Figure 2-1b). Most integrators and computer data systems list peak widths at half height as part of the report. Consult the instruction manual for the exact point of peak width measurement. The peak widths are reported in time units (usually minutes) since this is the form that most equations require.

Equations containing a peak width term have two forms - one using peak width at base and one using peak widths at half height. The equations using peak widths at the base can be found in Appendix A.

Figure 2-1: Peak Width
Measurements

2.2.2 Peak Symmetry

Most calculations assume the peaks are symmetrical triangles (they are actually Gaussian). Peak symmetry is determined by bisecting the peak through the apex. The width at 10% height is measured for each half (Figure 2-2). The width of the back half is divided into the width of the front half. A perfectly symmetrical peak has a symmetry factor of 1.00. Deviations of ±0.05 from 1.00 are not unusual. Larger deviations may be an indicator of peak overloading, a co-eluting peak, activity (tailing) or an injector problem.

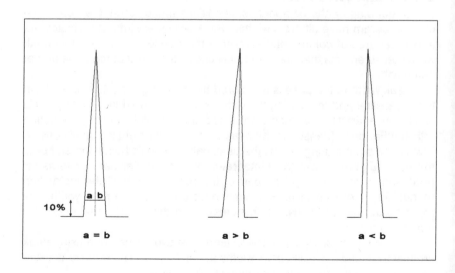

Figure 2-2: Peak Symmetry

2.3 Retention
2.3.1 Retention Time (t_r)
One of the most fundamental measurements in GC is peak (compound) retention. Retention time (t_r) in minutes is the usual unit of measure. It is the value obtained or reported by recording devices. The retention time is a measure of how long it takes a compound to travel down the column. It can also be considered the time a compound spends in the column. It is the sum of the time the compound spends in the stationary and mobile phases.

Retention times are dependent on the column, column temperature and carrier gas linear velocity. Due to small changes in the GC, retention times have a tendency to drift over time. It is also normal for retention times to vary by as much as ±0.05 minutes from run-to-run.

2.3.2 Adjusted Retention Time (t_r')
Compounds move down a column only while they are in the mobile phase (carrier gas). For a given column temperature and carrier gas linear velocity, every compound spends the same amount of time in the mobile phase. The time in the mobile phase is determined by injecting a non-retained compound and measuring its retention time (a list of non-retained compounds can be found in Table 6-1). This retention time is called the column dead time or hold up time and is denoted as t_m or sometimes as t_0.

Since every compound spends the same amount of time in the mobile phase, differences in compound retention are due to the differences in the times spent in the stationary phase. If the column dead time (t_m) is subtracted from a peak's retention time (t_r), a measure of the actual retention of a compound by the stationary phase is obtained. This value is called the adjusted retention time (t_r') and is calculated using Equation 2-1.

$$t_r' = t_r - t_m$$

t_r = retention time
t_m = retention time of non-retained peak

Equation 2-1: Adjusted Retention Time (t_r')

2.3.3 Partition Ratio (k)

Another retention measurement is called the partition ratio (k). In some cases, it may be called the capacity factor or be denoted as k'. The partition ratio is proportional to the time the compound spends in the stationary phase (t_r') relative to the time it spends in the mobile phase (t_m). Since all compounds spend the same amount of time in the mobile phase, k is a unitless measure of how long a compound spends in the stationary phase compared to another compound. The partition ratio is calculated using Equation 2-2. Compound retention increases as its k value increases. A compound is non-retained if k = 0.

$$ k = \frac{t_r - t_m}{t_m} $$

t_r = solute retention time
t_m = retention time of a non-retained peak

Equation 2-2:   Partition Ratio (k)

Figure 2-3:   Comparison of k and Retention Time

The partition ratio is a more direct measure of the actual magnitude of compound retention than retention time. A compound with k = 8 is twice as retained as a compound with k = 4; however, a compound with a retention time of 8 minutes is not twice as retained as a compound with a retention time of 4 minutes. Figure 2-3 shows an example of difference between k and retention time. Retention times can be misleading when measuring column retention. For this reason, partition ratios are used in many equations, theoretical discussions and quality control measurements.

2.3.4 Retention Index (I)

Retention index (I) is a measure of the relative retention of a compound relative to normal alkanes (hydrocarbons) at a given temperature on a particular stationary phase. Retention indices are also known as Kovats indices. Retention indices normalize instrument variables so that retention data can be effectively compared for different chromatographic systems. The retention index for a compound is calculated for isothermal temperature conditions using Equation 2-3.

Normal alkanes have retention indices that correspond to their number of carbons multiplied by 100. For example, n-$C_{12}H_{26}$ has I = 1200 and n-$C_{13}H_{28}$ has I = 1300. A compound with a retention index of 1225 elutes between n-$C_{12}H_{26}$ and n-$C_{13}H_{28}$, but closer to n-$C_{12}H_{26}$ under the same test conditions and column.

$$ I = 100\,y + 100\,(z-y) \left(\frac{\log t_r'(x) - \log t_r'(y)}{\log t_r'(z) - \log t_r'(y)} \right) $$

t_r' = adjusted retention times
x = compound of interest
y = normal alkane with y carbon atoms eluting **before** compound x
z = normal alkane with z carbon atoms eluting **after** compound x

Equation 2-3:   Retention Index (I) for Isothermal Conditions

Retention indices are useful when comparing relative elution orders of various compounds for a given column and conditions. Also, retention indices are good for comparing the retention behavior of two columns with the same description used under the same conditions.

For sample mixtures containing compounds with a wide range of boiling points, using isothermal conditions to determine retention indices is not practical. The retention times of the late eluting compounds are too long. Using Equation 2-3 to calculate retention indices for temperature program conditions results in errors especially if temperature programs starting at high temperatures, over a wide temperature range and with rapid ramp rates are used. Equation 2-4 is used to calculate retention indices for temperature program conditions. It is important to note that programmed indices require specification of column dimensions, carrier gas type and flow rate and the temperature program.

$$I_T = 100 \left(\frac{t_r(x) - t_r(y)}{t_r(z) - t_r(y)} \right) + y$$

t_r = retention times
x = compound of interest
y = normal alkane with y carbon atoms eluting *before* compound x
z = normal alkane with z carbon atoms eluting *after* compound x

Equation 2-4:   Retention Index (I_T) for Temperature Program Conditions

2.4 Phase Ratio (β)

The phase ratio (β) is a unitless value relating the diameter and film thickness of the column (Equation 2-5). Since β is unitless, the same units must be used for the column radius and film thickness (usually micrometers).

A table of phase ratios for various sized columns can be found in Appendix B. The phase ratio is used to determine the impact of column diameter and film thickness on retention (Section 2.5).

$$\beta = \frac{r}{2d_f}$$

r = column radius (μm)
d_f = film thickness (μm)

Equation 2-5: Phase Ratio (β)

2.5 Distribution Constant (K_D)

Upon entering the column, compound molecules are distributed between the mobile and stationary phases according to column temperature, the compound and stationary phase structures. The ratio of the concentrations of the compound in the stationary and mobile phases is defined as the distribution constant (K_D) (Equation 2-6). Figure 2-4 is a graphical representation of K_D.

$$K_D = \frac{\textit{Compound concentration (Stationary phase)}}{\textit{Compound concentration (Mobile phase)}}$$

Equation 2-6: Distribution Constant (K_D)

Figure 2-4: Graphical Representation of the Distribution Constant (K_D)

Due to the overwhelming number of possible temperatures, compounds and stationary phases, there are no comprehensive tables or graphs that list K_D values. The distribution constant is most useful in describing and predicting retention changes upon changing column dimensions and temperature.

The distribution constant can be related to retention, column diameter and stationary phase thickness using Equation 2-7. The phase ratio (β) depends only on column diameter and film thickness, and is fixed for a column; K_D is constant for a specific compound, stationary phase and column temperature. This means that K_D and β are independent and are affected by different variables. Also, note that column length, carrier gas type and flow rate are not directly involved in the terms in Equation 2-7.

$$K_D = k\,\beta$$

k = partition ratio
β = phase ratio

Equation 2-7: Distribution Constant (K_D) Relationship

2.5.1 K_D and Column Dimensions

For a given compound, stationary phase and column temperature, a change in column diameter or film thickness causes a change in the retention (k) of the compound. This occurs because the distribution constant remains constant for the same column temperature and stationary phase. The retention of the compound shifts depending on whether β increases or decreases. Table 2-1 summarizes the effects of changing the film thickness and diameter on retention (constant K_D). If both diameter and film thickness change, the direction of the retention shift depends on the relative change of the two dimensions.

Table 2-1:   Effects of Changing Column Diameter and Film Thickness on Retention

Column Diameter	Film Thickness	Phase Ratio	Retention
Increase	-	Increase	Decrease
Decrease	-	Decrease	Increase
-	Increase	Decrease	Increase
-	Decrease	Increase	Decrease

These relationship are valid only for constant column temperature, stationary phase and compound.

2.5.2 K_D and Column Temperature

Since the phase ratio (β) is fixed, retention (k) is affected by a change in K_D. Temperature is related to K_D by an inverse natural log (ln) function. This means that a small change in column temperature correspond to large change in retention. From $K_D = k\beta$, retention decreases as column temperature increases (i.e., decrease in K_D) and k increases as column temperature decreases (i.e., increase in K_D). The distribution constant does not change uniformly with temperature for all compounds. Upon a change in column temperature, the change of retention may be different for each peak in the chromatogram.

2.6   Column Efficiency

Symmetrical peaks with the smallest widths are desired in chromatography. Two peaks that are very narrow can be closer together than two broader peaks and still not overlap. All of the peaks in a chromatogram are not the same width. Ignoring injector influences on peak widths, peak widths increase as retention times increases. The effect is more obvious with isothermal temperature conditions. Column efficiency is the relationship between a peak's retention time and its width. This helps to account for the occurrence of broader peaks at longer retention times.

2.6.1 Number of Theoretical Plates (n)

Column efficiency is expressed as the number of theoretical plates (n). A peak in the chromatogram is used to calculate the number of plates using Equation 2-8. The number of plates per meter of column (n/m) is often reported so that columns of different lengths are more easily compared. A higher number of theoretical plates results in thinner peaks at their respective retention times. This is critical when attempting to separate two closely eluting peaks. A column with a low number of plates has more of each peak overlapping with each other.

$$n = 5.545 \left(\frac{t_r}{W_h} \right)^2$$

t_r = retention time of peak
W_h = peak width at half height (in the same units of time as t_r)

Equation 2-8:   Number of Theoretical Plates (n)

2.6.2 Height Equivalent to a Theoretical plate (h)

Another measure of column efficiency is the height equivalent of a theoretical plate (h or HETP). It is calculated using Equation 2-9. The shorter each theoretical plate, the greater the number that fits into an unit length of column, thus the greater the number of total theoretical plates per meter. High efficiency column (large n/m) have small values of h.

$$h = \frac{L}{n}$$

n = total number of theoretical plates
L = column length (mm)

Equation 2-9: Height Equivalent to a Theoretical Plate (h)

2.6.3 Effective Theoretical Plates (N) and Plate Heights (H)

A non-retained compound peak has a retention time and peak width; therefore, the theoretical plate number and plate height value are greater than zero. In theory, both values should be equal to zero. This problem is corrected by using the adjusted retention times (t_r') in Equation 2-7 to calculate n and Equation 2-9 to calculate h. The corresponding efficiency values when t_r' is used are called the effective theoretical plate number (N) and the height equivalent of an effective theoretical plate (H or HEETP). If the non-retained peak retention time is used, N = 0 and H = 0 are obtained. The error introduced by using retention time is small, thus often ignored.

2.6.4 Precautions When Using Theoretical Plates

Care must be exercised when comparing theoretical plate numbers. The number of plates is influenced by the retention of the peak used in the calculation. In theory, the calculated number of theoretical plates should be the same regardless of the retention of the peak used in the calculation. In reality, the number of theoretical plates is dependent on the retention of the peak used to calculate n (Figure 2-5). When calculating n values, a peak with k > 5 should be used to ensure valid and consistent numbers. Using a peak with a k < 5 inflates the number of theoretical plates.

Figure 2-5: Comparison of Theoretical Plate Numbers and Retention

Another important consideration is that theoretical plate numbers are actually measures of the performance of the GC system and not just the column. Other factors such as injector type and performance, and carrier gas flows can have significant influence on the plate numbers. Instrument settings and conditions need to be identical when comparing column plate numbers. This is especially true when comparing columns from different manufacturers. The k values for the compounds used in the calculation must be the same or the comparison is not valid. In most cases, each column manufacture use their own compounds and test conditions.

The use of temperature programming decreases peak widths. Theoretical plate numbers generated for the same column, but with different temperature conditions can not be compared. Depending on the temperature program, dramatically different plate numbers are obtained with each program (Figure 2-6).

Figure 2-6: Comparison of Theoretical Plate Numbers and Temperature Conditions
Compound: Methyl decanoate
Column: DB-1, 15 m x 0.25 mm I.D., 0.25 µm   He at 32 cm/sec
Isothermal: 100°C   *Program:*   60-140°C at 7°/min

2.7 Coating Efficiency (CE%)

Another slightly different measure of column efficiency is coating efficiency (CE) or utilization of theoretical plates (UTE). Coating efficiency is the comparison between a column's actual efficiency and its theoretical maximum efficiency (Equation 2-10). Equation 2-11 is often used to calculate the theoretical maximum efficiency (h_{min}).

$$CE\% = \left(\frac{h_{min}(actual)}{h_{min}(theory)} \right) \times 100$$

h_{min} = height equivalent to a theoretical plate

Equation 2-10:   Coating Efficiency (CE%)

$$h_{min}(theoretical) = r \sqrt{\frac{(11k^2 + 6k + 1)}{3(1 + k)^2}}$$

r   = column radius (mm)
k   = partition ratio

Equation 2-11: h_{min} theoretical

A coating efficiency of 100% means the efficiency is equivalent to that expected for a column with a stationary phase of uniform film thickness and without any physical or chemical flaws. Coating efficiencies less than 100% (which is normal) are departures from this perfect behavior. There is some error (less than ± 5%) in the calculation of theoretical values and experimental values, thus it is possible to occasionally obtain coating efficiencies slightly greater than 100%.

The maximum coating efficiency possible is dependent on the type of stationary phase. In general, non-polar stationary phases have higher coating efficiencies than polar phases. Typical coating efficiencies are 85-95% for non-polar phases and 60-80% for polar phases. Also, thicker film columns have lower coating efficiencies than the same phase with a thinner film.

Coating efficiency can be thought as a measure of the uniformity of the stationary phase. A more uniform phase results in a column of higher efficiency. Coating efficiency is not the amount of column coated with phase or the amount of column coated at the prescribed film thickness.

2.8 Separation Factor (α)

The separation factor (α) is a measure of the amount of peak separation. It is sometimes called selectivity. The separation factor is easily calculated using Equation 2-12 and is always equal to or greater than 1. If $\alpha = 1$, the peaks have the same retention, thus they completely co-elute. The separation factor does not contain any information about peak widths. It is only a measure of the distance (time) between two peaks.

$$\alpha = \frac{k_2}{k_1}$$

k_1 = partition ratio of the earlier eluting peak
k_2 = partition ratio of the later eluting peak

Equation 2-12: Separation Factor (α)

2.9 Resolution (R)

Resolution (R) is the measure of the amount of separation between two peaks taking the width of the peaks into account. Equation 2-13 is used to calculate the amount of resolution. A resolution of 1.5 represents fully resolved peaks without any baseline or space between the two peaks (Figure 2-7a). Resolution values greater than 1.5 represent peaks with some baseline between them (Figure 2-7b). Resolution values less than 1.5 represent peaks that have some degree of overlap or are partially resolved (Figure 2-7c).

$$R = 1.18 \left(\frac{t_{r2} - t_{r1}}{W_{h1} + W_{h2}} \right)$$

t_{r1} = retention time of peak 1
t_{r2} = retention time of peak 2
W_{h1} = peak width at half height of peak 1 (same time units as t_r)
W_{h2} = peak width at half height of peak 2 (same time units as t_r)

Equation 2-13:   Resolution (R)

Resolution is often given as a percentage value. Resolution expressed as a percentage is useful when peaks are not completely resolved. Less than 100% resolution is easy to visualize since the percentage value describes the size of the valley between the two peaks relative to the height of the smallest peak (Figure 2-7c). One problem with resolution expressed as a percent is the inability to numerically distinguish between barely baseline and much greater than barely baseline separations. Both situations will be represented as 100% resolution even though the one set of peaks is separated by a substantially greater amount (Figures 2-7a and b).

Peak separation (α) means little if peak resolution is not considered. Two broad peaks can have equal or better separation than two narrow peaks, but the narrow peaks are better resolved (Figure 2-8a and b). Less separation can be tolerated if the two peaks are narrow. The two peaks can be closer together and still maintain adequate resolution (Figure 2-8c).

Figure 2-7:   Resolution Numbers

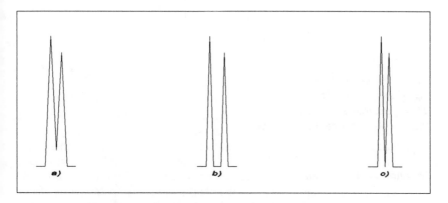

Figure 2-8:   Separation versus Resolution
a) Separation but incomplete resolution
b) Equal separation but better resolution
c) Less separation but adequate resolution (better than in a)

In chromatography, the real goal is peak resolution. If resolution is achieved, separation automatically occurs. Instrument setup and operation, and column selection should be primarily focussed on obtaining good resolution and less on obtaining separation. For low efficiency situations, more separation is needed to resolve the peaks. For high efficiency situations, less separation is needed to resolve the peaks.

2.10   Trennzahl (TZ)

A different measure of resolution is Trennzahl (TZ) or separation number. Like resolution it takes peak widths and the amount of separation into account. Two consecutive n-alkanes (hydrocarbons) are used as the peaks. TZ is calculated using Equation 2-14. TZ can be interpreted as the number of peaks (of equal width) than can be placed side by side between the two peaks. The higher the number, the greater the resolution for the two peaks. The TZ number is dependent on the compounds (peaks) used; therefore, the identity of the two hydrocarbons needs to be specified along with the TZ value.

$$TZ = \left(\frac{t_{r2} - t_{r1}}{W_{h1} + W_{h2}} \right) - 1$$

t_{r1} = retention time of peak 1
t_{r2} = retention time of peak 2
W_{h1} = peak width at half height of peak 1
W_{h2} = peak width at half height of peak 2

Equation 2-14:   Trennzahl

2.11 Column Capacity

Column capacity is the amount of compound that can be introduced into the column without causing excessive peak shape distortion. When too much of a compound is placed in the column, its peak begins to develop a sloping front and may appear to be asymmetrical. A peak is overloaded when it has a symmetry factor of 1.10 or greater (Figure 2-10a). A peak with a symmetry factor slightly above 1.10 often does not appear to be asymmetrical or overloaded. Often peak overloading is determined by visual inspection. A shark fin shape is the most common description and appearance of an overloaded peak (Figure 2-10b).

Column capacities are usually quoted as per peak and not for the entire sample. For example, a column capacity of 150 ng refers to 150 ng per compound or peak and not the mass of the injected sample. Column capacity does not refer to the volume of sample. The maximum sample volume is primarily a function of the injector, column dimensions, sample solvent and initial column temperature. It is not unusual to have a column's capacity differ by 2-3 times for dissimilar compounds .

Greater column capacity is obtained when the polarity of the stationary phase and compound are similar. Larger diameter and thicker film columns have greater capacities. Column capacity decreases as compound retention increases. At the same concentration, the early eluting peaks in a chromatogram have satisfactory shape while the later eluting peaks exhibit overloading characteristics.

A column is not harmed by injecting too much sample. The primary problem is the asymmetric peak. It is visually unappealing and it may not be integrated properly by the data system.

Figure 2-10: Overloaded Peak

3   CAPILLARY GC COLUMNS: TUBING

3.1 Fused Silica Capillary Columns

Capillary columns can be thought of having three distinct "layers" or parts. They are the fused silica tube, the protective coating and the stationary phase (Figure 3-1). It is assumed that "fused silica tubing" refers to tubing with the protective coating.   The stationary phase is responsible for compound retention and separation (Chapter 4). Fused silica tubing without stationary phase is available.

Figure 3-1:   Cross Section of a
           Fused Silica Capillary
           Column (Not to Scale)

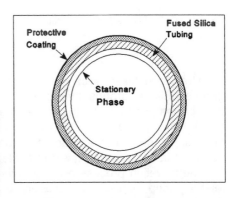

3.2 Fused Silica Tubing

The only interaction occurring inside a capillary column should be between the stationary phase and the sample compounds. This requires an inert inner surface that does   not   interact   with   the compounds. Tubing that is not inert is called active, thus it exhibits activity. Fused silica is the material of choice for capillary column tubing since its purity is unsurpassed by any other suitable glasses. Fused silica is a synthetic, quartz-like glass containing typically less than 1 ppm of metallic oxide impurities. A very low concentration of these oxides is required to minimize the reactivity of the fused silica surface. Tubing made of borosilicate glass should be avoided if possible.

Figure 3-2:   Silanols and Interactions

Untreated fused silica is very active due to the presence of silanol groups. Silanol groups interact with compounds with hydrogen bonding capabilities (Figure 3-2). In general, these types of compounds have thiol (-SH), hydroxyl (-OH), amine (-NH) or carboxylic acid (-COOH) functional groups. Interaction with an active surface results in tailing or reduced size peaks. Fused silica tubing is deactivated by the use of silylating reagents. The deactivation process eliminates the silanols by converting them into relatively nonpolar and nonreactive groups (Figure 3-3). Even after the most rigorous deactivation process, some surface silanols are still present and contribute to the residual activity of columns. Minimizing the number of accessible silanol groups is the realistic goal of surface deactivation.

Columns lose some of their inertness with continued use. Slow loss of the deactivating groups with conversion back to the silanols occurs. Also, normal stationary phase degradation with column use creates active sites in the stationary phase. These processes are not easily reversed. There are column treatment solutions that are available, but they have little to no permanent effect.

Figure 3-3: Deactivated Silanol
Example of a trimethylsilyl (TMS) deactivating group. Other types of deactivating groups may be used.

The deactivation process creates a suitable surface for the stationary phase. The stationary phase has to evenly coat the surface of the tubing. The most stable and robust deactivation processes create a relatively non-polar surface, thus non-polar stationary phases coat the tubing surface more evenly. For this reason, non-polar stationary phase columns are usually more efficient than more polar stationary phase columns of the same dimensions.

3.3 Outer Coating

Fused silica tubing is fairly fragile and requires some type of protective coating. The most common coating material is polyimide. This polymer coating serves a dual purpose. It fills flaws in the tubing which prevents the further growth of any defects, and it acts as a barrier to prevent corrosion and damage to the fused silica tubing. Polyimide coated columns are very strong and do not require delicate handling. Any handling or process causing scratches or abrasions in the polyimide coating results in a weak spot with possible breakage at a later time.

Standard polyimide coating is stable for only several days at 380°C. Longer exposure times or higher temperatures makes the polyimide begin to discolor, flake and peel away from the tubing. The tubing usually breaks at these locations. The polyimide coating has excellent stability at temperatures of 350°C or below. Recently, tubing with polyimide coatings stable to 400°C has become available. This tubing is usually found on high temperature columns intended to be regularly used at 350-400°C.

The polyimide coating is responsible for the brown color of fused silica tubing. The intensity of the brown color is influenced by slight batch-to-batch differences and the temperture of the polyimide coating process. The color has no effect on the performance of a capillary column. The coating is on the outside of the tubing while the separations occur inside the tubing. Some slight tubing strength or flexibility differences may be experienced with different color columns; however, there appears to be no pattern to the color and strength relationship. Column colors may range from very dark brown to a very light yellow. The color may darken with use especially if the column is routinely held at higher temperature.

Aluminum coated ("clad") fused silica tubing is also used. This tubing has a much higher upper temperature limit (~450°C) than polyimide coated tubing. One drawback to aluminum clad tubing is its brittleness when exposed to continuous changes in temperature. It is satisfactory for isothermal temperature work; however, the tubing becomes weak upon cycling of the oven temperature during temperature programming situations. This has precluded widespread use of aluminum as a coating material.

3.4 Other Tubing Materials

Recently, stainless steel clad columns have become available. This tubing is probably better described as stainless steel tubing with a thin, interior coating of fused silica. This tubing exhibits slightly higher activity than polyimide coated, fused silica tubing. This limits its use for compounds with little to moderate activity. Cracks in the fused silica can develop exposing the compounds to bare metal surfaces. The exposed metal surfaces are very active. An increase in activity problems may occur with continued use of the column. The metal coated tubing is not a flexible as polyimide coated tubing which makes installation slightly more difficult. Also, the metal coated tubing is more difficult to properly cut. The primary advantages of this tubing are its high temperature limits and ruggedness. This tubing should not be confused with stainless steel tubing.

Stainless steel columns have the stationary phase coated directly onto the metal tubing without the benefit of a glass or fused silica surface. These columns exhibit high activity due to the metal surface. Originally, capillary columns were made out of metal tubing. Recent advances in deactivation techniques have caused the reintroduction of stainless steel tubing. The newer stainless steel columns are superior to the original ones of 25-30 years ago; however, they are still not as inert as fused silica tubing. The advantages of stainless steel tubing are its very high temperature limits and ruggedness. The disadvantages are the higher activity, inflexibility and greater difficulty in properly cutting the tubing.

A few borosilicate glass columns are still around. They exhibit all of the characteristics that gave capillary columns their early, bad reputation. Borosilicate glass columns are very fragile, inflexible, very active and the stationary phases are easily disrupted.

4   CAPILLARY GC COLUMNS: STATIONARY PHASES

4.1 Stationary Phases

Capillary column stationary phases are polymers deposited on the inner walls of the tubing in a thin, uniform film. The polymers must withstand high temperatures for prolonged periods of time, be non-reactive, and evenly coat the surface of the tubing. The uniformity, thickness and chemical nature of the stationary phase are very important to the overall behavior and performance of a column.

The structure of the stationary phase influences the *separation* of the compounds. Column dimensions primarily affects the *resolution*. A column with very high resolution is not useful if the stationary phase can not separate the desired compounds. Capillary columns require relatively few stationary phases to achieve the separations necessary for complex samples. For this reason, only 12-15 distinct stationary phases are available for capillary columns.

When the separation of two compounds does not occur with one stationary phase, it is very likely that separation can be achieved with a different stationary phase. A stationary phase is selected primarily based on whether it provides adequate compound separation.   There are other considerations and limitations involved with stationary phase selection; however, these are relatively minor when compared to obtaining the required separations.

Unfortunately, there are no simple tables or foolproof formulas that can be used to easily select the proper stationary phase. On the other hand, there are some relatively simple concepts that can be successfully used in most cases. A basic understanding of the structure of common stationary phases and the properties of various compound functional groups is all that is necessary.

4.2 Types of Stationary Phases
4.2.1 Polysiloxanes or Silicones

The predominant polymers used as stationary phases are the substituted polysiloxanes. These stationary phases are sometimes referred to by their common name of silicones. Polysiloxanes are considered to be the stationary phases most resistant to abuse and they have superior lifetimes compared to most other types of stationary phases. Polysiloxanes are distinguished by their alternating silicon and oxygen linear backbone and two functional groups attached to each silicon (Figure 4-1). It is the structure of the functional groups and amount of each group that distinguishes each stationary phase.

Figure 4-1:   Base Siloxane
               Backbone Unit

There are only a few structures used as functional groups. They are limited to ones that can be attached to the polysiloxane backbone without significantly impairing the thermal stability of the polymer. Also, each group needs to change the stationary phase in a manner to make it have unique separation characteristics. The primary groups used are methyl, phenyl, cyanopropyl and trifluoropropyl. These groups are used in various amounts and combinations to impart specific separation characteristics to each stationary phase. The functional groups are not added to the backbone in a irregular pattern, but in a semi-ordered fashion. Most substitutions are in blocks of the same group then another block of the other group. The total numbers of each block is precisely regulated to maintain the bulk properties of the polymer.

The most elementary polysiloxane is the 100% methyl substituted. Most of the other polysiloxane stationary phases are mixtures containing a substantial portion of methyl. The most common mixtures are phenyl-methyl, cyanopropyl-phenyl-methyl, cyanopropyl-methyl, and trifluoropropyl-methyl. The base structures of these stationary phases are illustrated in Figure 4-2.

Figure 4-2:   Common Substituted Polysiloxanes

Stationary phase descriptions provide information about the type and amount of substitution on the polysiloxane backbone. Manufacturer's brand names often have no real meaning or significance and are not a reliable method to obtain structural information about a stationary phase. There are plenty of examples where similar brand name columns are not remotely equivalent in structure, and hence separation behavior.

Polysiloxane stationary phases use a percentage substitution system to designate stationary phase structures. For example, a 5% phenylmethylpolysiloxane is comprised of a phenyl group attached to 5%, by number, of the sites on the backbone silicon atoms; the remaining 95% of the silicon sites have methyl groups. Sometimes a stationary phase is named using a shorthand system. The previous example could be called a 5% phenyl phase. When not stated, the remaining group is methyl and at a percentage to make the total 100%.

The naming of cyanopropylphenyl polysiloxanes can be confusing. The cyanopropyl and phenyl groups are separate, but attached to the same silicon atom; the cyanopropyl group is not attached to the phenyl ring. For example, a 6% cyanopropylphenylmethyl polysiloxane has a cyanopropyl attached to 3% of the silicon sites, a phenyl on 3% of the sites (on the same silicon atoms as the cyanopropyl groups) and a methyl on 94% of the sites. Realizing that there are only four major functional groups (i.e., methyl, phenyl, cyanopropyl and trifluoropropyl) helps to clarify most stationary phase naming confusions.

Approximately 1% of the silicon backbone atoms have a vinyl group used for bonding and cross-linking (Section 4.3.1). In most cases, the 1% vinyl content is not stated. The carbon-carbon bonds comprising the bonding and cross-linking groups do not significantly influence the separation characteristics of the stationary phase, thus they are often ignored.

4.2.2 Polyethylene Glycols

Another common stationary phase is polyethylene glycol (PEG) (Figure 4-3). Polyethylene glycol stationary phases differ primarily by the molecular weight or chain length of the polymer chains. Carbowax 20M is probably the most widely used form of polyethylene glycol for capillary column stationary phases.

Figure 4-3: Polyethylene Glycol

$$HO-[CH_2-O-CH_2-O]_n-H$$

PEG stationary phases have some unique separation characteristics not exhibited by polysiloxanes. The major disadvantage of PEG stationary phases is their extreme sensitivity towards oxygen, especially at high temperatures. The presence of oxygen in the carrier gas causes rapid destruction of most capillary column stationary phases; polyethylene glycols are among the stationary phases most susceptible to this problem. PEG stationary phases have narrow temperature ranges where they can be safely used without damage or performance problems. The liabilities are tolerated since PEG stationary phases have been widely used for a long time, and there is a large database of separation data available in the published literature.

When analyzing samples that are acidic or alkaline, a PEG stationary phase may begin to mimic some of the pH characteristics of the sample. For example, if a large number of amine samples have been injected into the column, it may become slightly alkaline. Injections of an acidic compound results in a very poorly formed peak or no peak at all. This pH shift is reversible by solvent rinsing the column (only if it is a bonded and cross-linked phase) with a small amount of water, followed by acetone then hexane.

Acidic or alkaline PEG columns are available. They are used when non-tailing peaks are required for organic acids or bases that are separated with a PEG column. Obviously, an acidic column excludes the use of the column for alkaline compounds and vice versa. Unfortunately, these columns are not very stable and require frequent replacement.

4.2.3 Porous Layer Stationary Phases

A unique set of stationary phases are the porous polymers. Capillary columns with these stationary phases have a layer of small, porous particles coated onto the inner wall of the fused silica tubing. The particles are usually held in place by some type of chemical binder. The particles are not bound directly to the tubing, but only held in place by the binder. These types of columns are called porous layer open tubular or PLOT columns. Stationary phases of aluminum oxide (Alumina), molecular sieves and a series of porous polymers (Poraplot-like) are currently available.

For PLOT columns, gas-solid adsorption is the primary separation mechanism involved rather than gas-liquid partitioning as for polysiloxanes and PEG. The stationary phases used in PLOT columns are more retentive than liquid stationary phases. This makes PLOT columns well suited for the analysis of light hydrocarbons, sulfur gases, permanent gases or other very volatile compounds. Polysiloxanes require column temperatures well below 35°C to separate these types of compounds, thus necessitates the use of liquid nitrogen to cool the GC oven. This adds expense, complexity and time to the analysis. Many PLOT columns can separate these very volatile compounds at column temperatures of 35°C or above.

Since PLOT columns are so retentive, they are not suited for many types of compounds. Unless the compound is a gas at room temperature or a very low boiling liquid, the compound requires a very long time to exit the column; some may not exit the column at all. Peak width increases very rapidly with retention with PLOT columns. Beyond a k of about 10, the peak widths are very wide. PLOT columns are less efficient than columns with liquid stationary phases and are usually limited to wider diameter fused silica tubing. Depending on the stationary phase and conditions, some stationary phase particles can become dislodged from the column and possibly foul the GC detector.

4.3 Characteristics of Stationary Phases

4.3.1 Bonded and Cross-linked Stationary Phases

Long chain length polymers are used for stationary phases. To improve the stability and longevity of the stationary phases, the individual chains are connected together using a two carbon linking unit (Figure 4-4). This process is called cross-linking. Upon linking, the stationary phase is like a very large, branched polymeric molecule. This branched polymer is also attached to the surface of the fused silica tubing using the same type of two carbon unit linkage (Figure 4-4). This anchors the stationary phase to the tubing and

Figure 4-4: Bonded and Cross-linked Stationary Phase

prevents the stationary phase from leaving the column. This process is called bonding.

Before the advent of bonded and cross-linked stationary phases, capillary columns required substantial maintenance and were subject to some limitations in use. Some solvents could not be injected into the column because the stationary phase could be damaged. This usually caused poor peak shape and loss of column efficiency. Organic solvents and water do not damage bonded and cross-linked stationary phases. Peak shape problems occur with some compound, stationary phase and solvent combinations, but stationary phase damage does not occur. An added benefit of bonded and cross-linked stationary phases is that the column can be washed with solvents to remove contaminants. This substantially prolongs the useable life of the column and minimizes the potential damage caused by dirty samples.

Whenever possible, a column with a bonded and cross-linked stationary phase should be used. The long term stability, reproducibility, and performance are better. A few stationary phases are not available in bonded form. In these cases, greater attention to column care and maintenance is required to maximize column life and performance. The use of a retention gap is strongly recommended to minimize the amount of problems encountered with non-bonded stationary phases.

4.3.2 Stationary Phase Polarity

Stationary phase polarity is a bulk property of the polymer comprising the stationary phase. Stationary phase polarity is directly related to the amount and polarity of each functional group. The polarity of a stationary phase increases as the polarity of the substituted groups and their relative amounts increase. Table 4-1 lists the most common capillary column stationary phases in the order of increasing polarity.

Stationary phase polarity influences compound separation; however, stationary phase selectivity (Section 4.3.3) is a more important factor when considering separations. Stationary phase polarity should be considered when temperature ranges, column lifetime, capacity and bleed are important. It is common for more than one stationary phase to provide the desired separations. In these cases, use the most non-polar stationary phase. Non-polar stationary phases are superior to polar phases in several areas. Non-polar stationary phase columns can be used over a wider temperature range. For very low or high temperature analyses, stationary phase selection is primarily limited to the non-polar ones. Polar stationary phases are more

Table 4-1:   Stationary Phase Polarities

In Order of Increasing Polarity*

100% Methyl
5% Phenyl
6% Cyanpropylphenyl
20% Phenyl
35% Phenyl
14% Cyanopropylphenyl
50% Phenyl
65% Phenyl
40-50% Trifluoropropyl
50% Cyanopropylphenyl
Polyethylene glycol (PEG)
50% Cyanopropyl
80% Cyanopropyl
100% Cyanopropyl

* As based on Kovats Indices

fragile than non-polar phases. Under equivalent conditions, a non-polar stationary phase column has a longer lifetime. Polar stationary phases generate a higher background signal (bleed) than non-polar stationary phases. This makes non-polar stationary phase columns better for low level analysis of compound that elute at high temperatures. Greater capacity is obtained for stationary phases close in polarity to the compounds. For example, a column with a polar stationary phase has greater capacity for polar compounds than non-polar compounds.

Using a more polar phase never guarantees the separation of compounds that co-elute on a less polar phase. Polar columns are often tried as a means to improve a separation in the mistaken belief that increased polarity translates into better separation. This approach is successful on occasion, but due to other factors not directly related to the change in stationary phase polarity.

4.3.3 Stationary Phase Selectivity

For two compounds to be separated, each must spend a different amount of time in the column. This is accomplished when one compound spends more time in the stationary phase than the other. The time spent in the stationary phase depends on the interaction of each compound with the stationary phase. These interactions affect retention, and thus also compound separation.

The interactions are dependent on the structure of the compounds and the stationary phase. Each stationary phase has a particular set of interactions that it can undergo with the compounds. If a compound has one of the characteristics required to participate in the interactions, the separation is influenced by those interactions. If a compound does not have a characteristic required for a particular interaction, that interaction does not occur and does not have an influence on the separation. This behavior is called stationary phase selectivity. The stationary phase selectively interacts with some compounds at the exclusion of others.

4.3.4 Stationary Phase Interactions

There are three major interactions associated with the liquid stationary phases used in capillary columns. Stationary phases have one or more of these interactions depending on its structure. The three interactions are dispersion, dipole and hydrogen bonding. In order to discuss selectivity on a practical level, some over-simplifications and generalization have to be made. Instead of describing the actual interactions on a molecular level, retention and separation behavior are described using common physical and chemical properties. These simplifications are valid in most cases.

4.3.4.1 Dispersion Interaction

Dispersion is the most basic and universal of the stationary phase interactions. As shown in Table 4-2, all stationary phases exhibit a strong dispersion interaction. Dispersion is the primary separation mechanism for all compounds. The dispersion interaction is related to the intermolecular attraction between the compound and stationary phase. The polarizability of the compound plays a major role in this interaction. Larger molecules are more polarizable than smaller molecules of similar functionality.

An erroneous, but common simplification for dispersion-only stationary phases is to relate retention to compound boiling points. That is, compounds with higher boiling points are more strongly retained. This can be a useful simplification in some cases (e.g., compounds in a homologous series); however, there are many examples that do not follow this simplification (e.g., a collection of compounds with a mixture of functional groups and structures). A better simplification is to relate retention to compound vapor pressures. Compounds with lower vapor pressures at the column temperature are more strongly retained on dispersion-only stationary phases. The vapor pressure simplification still involves some error, but it is much smaller than using compound boiling points. Unfortunately, compound vapor pressures are not as readily available in references sources as compound boiling points.

The solubility of the compound in the stationary phase is one aspect of dispersion. Higher compound solubility in the stationary phase results in greater retention. Compound solubilities in stationary phases can be difficult to estimate. Polar compounds are more soluble (i.e., more retained) in a polar stationary phase than a non-polar stationary phase; non-polar compounds are more soluble in a non-polar stationary phase than a polar stationary phase. Difficulties arise when trying to determine the retention compounds of diverse polarities and structures. For example, the relative retention of a large non-polar compound and a small polar compound on a polar stationary phase can be difficult to predict. The small size of the polar compound results in low retention while its polar character results in high retention on the polar stationary phase; the large size of the non-polar compound results in high retention while its non-polar character results in low retention on the polar stationary phase. It is not easy to determine whether compound size or polarity has the greatest influence on retention with this stationary phase.

Table 4-2:   Dispersion Interaction

Functional Group	Dispersion Interaction Strength
Methyl	Strong
Phenyl	Very Strong
Cyanopropyl	Strong
Trifluoropropyl	Strong
PEG	Strong

Interaction strengths: None, Weak, Moderate, Strong, Very Strong

4.3.4.2 Dipole Interaction

The dipole interaction is related to the interaction between the stationary phase and compounds with dipole moments. Cyanopropyl substituted polysiloxanes and PEG stationary phases have particularly strong dipole interactions (Table 4-3). If a compound does not have a dipole moment (permanent or induced), the dipole interaction does not have an influence on retention. Also, if a compound has a dipole, but the stationary phase does not exhibit a dipole interaction, a dipole interaction does not occur.

A compound has an appreciable dipole if there is uneven charge distribution due to the structure of the compound. This does not mean that the compound has a charged group. As long as part of the molecule is slightly positive and another is slightly negative, the compound has a dipole moment. There are some generalizations that aid in determining whether compounds have moderate to large dipole moments. Compounds with heteroatoms (O, N, S, P) or halogens (Cl, F, Br, I) usually possess a dipole moment. Also, compounds with hydroxyl (-OH) and amine (-NH) functional groups usually have large dipoles. Compounds with double bonds, ethers, ketones, and many esters usually have moderate to small dipole moments. Saturated hydrocarbons and compounds that are completely symmetrical (even if they have heteroatoms or halogens) usually have very to zero dipoles. For compounds with complex structures or multiple functional groups, it can sometimes be difficult to determine if there is a significant dipole moment. Dipole moments are more dependent on the overall structure of the compounds and not on the functional groups themselves. There is a limited amount of dipole information available in the literature, but it is usually limited to common and widely used compounds.

It is the difference in the dipole moments between the compounds rather than the size of their dipoles that is most important. Regardless of the strength of the dipoles, if they are the same for two compounds, their dipole interactions are the same. If two compounds have different dipoles, the different dipole interactions may lead to different retention (ignoring the other interactions).

If the dipole difference between two compounds is small, a stationary phase with a large dipole interaction is needed. A large amount of the dipole interaction characteristic is necessary to distinguish the small difference between the two compound's dipole moments. For example, a PEG stationary phase or one with a high cyanopropyl content is required to separate compounds with small dipole differences that do not separate on other types of stationary phases.

Table 4-3:   Dipole Interaction

Functional Group	Dipole Interaction Strength
Methyl	None
Phenyl	None
Cyanopropyl	Very Strong
Trifluoropropyl	Moderate
PEG	Strong

Interaction strengths: None, Weak, Moderate, Strong, Very Strong

4.3.4.3 Hydrogen Bonding Interaction

The hydrogen bonding interaction is related to the ability of the stationary phase and compound to hydrogen bond. It is not overly important which is the proton acceptor or donor as long as the interaction occurs. Table 4-4 shows that the strongest hydrogen bonding interaction occurs for the cyanopropyl and PEG stationary phases.

In general, a compound exhibits strong hydrogen bonding if it contains a hydroxyl (-OH) or amine (-NH) group. Other functional groups such as esters, ethers and ketones have moderate to small hydrogen bond strengths. Alkyl groups (hydrocarbons) and halogens have weak to no hydrogen bonding strength. Hydrogen bonding is more dependent on the actual functional groups and not on the overall structure of the compound. There is little information in the literature on hydrogen bonding strengths.

It is the difference in the hydrogen bonding strength of the compounds and not the absolute strength of the bonding that determines the result of the interaction. Ignoring all other interactions, if two compounds have different hydrogen bonding strengths with the stationary phase, differences in retention may be observed. The smaller the difference in the hydrogen bonding strengths, the greater the hydrogen bonding interaction needed by the stationary phase. A PEG stationary phase or one with high cyanopropyl content is required to separate compounds with small hydrogen bonding differences that do not separate on other types of stationary phases.

Table 4-4:   Hydrogen Bonding Interaction

Functional Group	Hydrogen Bonding Interaction Strength
Methyl	None
Phenyl	Weak
Cyanopropyl	Moderate
Trifluoropropyl	Weak
PEG	Moderate

Interaction strengths: None, Weak, Moderate, Strong, Very Strong

4.3.4.4 When There are Multiple Interactions

Many polysiloxane stationary phases exhibit multiple interactions due to their structure (Table 4-5). The influence of each interaction on the separations is difficult to predict. One interaction may be stronger than another and each interaction affects each compound to a different extent. The amount of each functional group influences the overall interaction properties of the stationary phase. For example, both cyanopropyl and PEG stationary phases have strong dipole and hydrogen bonding interactions. PEG stationary phases are usually 100% polyethylene glycol while the cyanopropyl content is usually less than 100% for most polysiloxanes. The dipole and hydrogen bonding interactions are stronger for the PEG stationary phase and have a greater impact on the separations.

Table 4-5:   Stationary Phase Interactions

Functional Group	Dispersion	Dipole	Hydrogen Bonding
Methyl	Strong	None	None
Phenyl	Very Strong	None	Weak
Cyanopropyl	Strong	Very Strong	Moderate
Trifluoropropyl	Strong	Moderate	Weak
PEG	Strong	Strong	Moderate

Interaction strengths: None, Weak, Moderate, Strong, Very Strong

Another problem with predictions about separations is the conflicting effects of two different interactions. One interaction may move the compound peaks further apart while the other has the effect of moving them closer together. For example, two compounds may separate due to the dispersive interaction, but the dipole interaction may alter the relative retention enough to cause co-elution.

Generally, examining the interactions is useful only after a chromatogram is obtained for a particular column. If there is a separation problem that can not be solved by changing conditions, changing to a different stationary phase is usually the next step. Success in finding a new stationary phase is improved by examining the compound and stationary phase interactions. At least certain stationary phases can be eliminated based on the lack of the necessary interactions or absence of significant change in the interactions. In some cases, there may be more than one stationary phase that provides adequate separation.

4.4 Stationary Phase Equivalencies

Most column manufacturers control the synthesis of their stationary phases from start to finish, and they deactivate the fused silica tubing using proprietary processes. This control results in high quality and reproducible columns, but ones that are slightly different. A column from one manufacturer may not be exactly the same as an equivalent stationary phase and size column from another supplier. In most cases, the differences are insignificant. Performance differences are more pronounced with polar stationary phases. The most common differences are the degree of inertness and slight retention variations. Quality differences are evident with efficiency, bleed and reproducibility being the main distinguishing characteristics.

Particular care must be taken when comparing specialty phases. These are columns designed for specific analyses and the stationary phase structures are usually proprietary. The stationary phases may be different for columns with apparently similar brand names (e.g., DB-608 and SPB-608). This can make specialty column comparisons difficult since the stationary phases may be different, but their structures are unknown. The separations obtained with these columns should be the guide when determining the equivalencies of specialty columns. Each column may have significantly different separations, making one column a better choice for a specific analysis or conditions.

4.5 Column Temperature Limits

All stationary phases have a temperature range over which they are stable and functional for extended periods. This range usually defines the temperature limits for a capillary column. Exceeding the temperature limits may lead to permanent damage to the column and loss of performance.

Below the lower temperature limit, the stationary phase loses its chromatographic properties. If the column is below its lower temperature limit, the peaks become broad and rounded. The problems becomes worse the further the column temperature is below its limit. The peaks regain their narrow and sharp form when the column is heated above the lower temperature limit. Permanent damage does not result if the column is exposed to temperatures below its lower limit. Cryocooling a column to focus a sample at the front of the column does not damage a column. Separation characteristics return as soon as the column heats up beyond its lower limit. At temperature below -60°C, some brittleness of the fused silica tubing may be experienced.

The upper temperature limit is usually given as two numbers (e.g., 325/350°C). In some cases, there may only be one upper limit. The first temperature is called the isothermal limit. This is the highest temperature that the column can be exposed to for indefinite periods of time. The second temperature is called the temperature program limit. The column can be left at this temperature for 10-15 minutes.

Exceeding the upper temperature limits results in damage to the stationary phase. Slight overexposure of the column to excessive temperatures does not result in the instant destruction of the column. In most cases, a small amount of column lifetime is sacrificed. The amount of lifetime loss is dependent on the temperature, the duration at the elevated temperature, the stationary phase, and the amount of oxygen present in the carrier gas. Greater damage occurs with longer exposure times, larger temperature excesses, higher oxygen concentrations, and a lack of carrier gas flow. Polar stationary phases are more susceptible to damage and degrade at a faster rate when exposed to excessively high temperatures.

Thermal damage is evident as excessive column bleed and peak tailing for active compounds. Loss of retention and efficiency (resolution) may be evident for severe cases of thermal damage. Temperature limits do not have any extra safety factors built into the values. Maintaining the column below its upper temperature limits should be practiced to ensure reasonable column life and performance.

Column temperature limits may differ slightly between manufacturers. Depending on column lifetime and especially bleed criteria, the limits may be set 10-20°C lower or higher. Be careful to distinguish between a high temperature version of a stationary phase and the regular version. The most common temperature limits for a variety of stationary phases can be found in Appendix C.

4.6 Column Bleed
4.6.1 What is Column Bleed?

Column bleed is one of the most misunderstood aspects of capillary columns. Column bleed is evident as a rising baseline at temperatures near the upper temperature limit of the column. The rising baseline is caused by the elution of the degradation products of a column's stationary phase. This degradation is normal and increases at elevated temperatures. The degradation products are always present and are not necessarily a sign of a damaged column. Every column regardless of the source or quality exhibits some stationary phase bleed.

The primarily degradation process is breakage of the polymer chain at the Si-O bond. The amount of bond cleavage increases at higher temperatures. If the breakage occurs near the end of a polymer chain, the resulting small fragment is volatile enough to elute from the column. It is the continuous elution of these fragments that is responsible for column bleed. The sharp rise in the baseline is caused by the increasing amount of degradation species as the temperature increases. Oxygen acts to increase the amount of bond breakage, thus accelerating the rate and amount of stationary phase degradation. In general, polar stationary phases as less stable, thus their bleed levels are higher. Temperature limits are adjusted to partially accommodate for this behavior

Stationary phase degradation is temperature dependent. The rate and amount of degradation product formation is constant under isothermal column conditions. This means that the baseline should be flat at constant temperature. Any rise or deviation in the baseline during constant temperature situations is not due to column bleed, but to another source. In most cases, the baseline deviation is caused by system contamination.

Stationary phase degradation increases at higher column temperatures. If a column is temperature programmed, the degradation increases as the temperature program progresses. At lower temperature ranges, the baseline rise is small or not evident since the amount of stationary phase degradation is negligible. As the column temperature approaches the column's upper temperature limit, the amount of degradation becomes much greater. The degradation starts to rapidly increase at 30-40°C below the isothermal temperature limit of the column, thus a sharply rising baseline is seen as the temperature program passes through this temperature region.

The appearance of column bleed is best illustrated by a bleed profile (Figure 4-5). A bleed profile is generated by running a temperature program without an injection (i.e., blank run). The temperature program should follow these guidelines: initial temperature of 50-100°C, ramp rate of 10-20°C, final temperature equal to the isothermal temperature limit of the column, and a final hold time of 10 minutes. The data system or integrator should be set so that the size of the baseline rise is 25-50% of full scale. All of the features of column bleed are evident in the bleed profile:

1) No to little rise in the baseline at the lower temperature regions
2) Sharp rise starting at 30-40°C below the upper temperature limit
3) Flat baseline during isothermal temperature conditions

Figure 4-5: Bleed profile
Column: DB-1, 15 m x 0.32 mm I.D., 0.25 µm
Oven: 100-325°C at 15°/min, 325°C for 10 min
Baseline rise equivalent to the height of a 2 ng n-C_{14} peak

It is important to notice the absence of peaks in the bleed profile. A column can not generate any peaks by itself. Introduction of a compound into the column followed by eventual elution from the column is required to obtain a peak. Bleed species originate along the entire length of column and elute from the column in a continuous stream. Bleed is a continuous process, thus a continuous signal is generated at the detector. This is the reason that column bleed is observed as part of the baseline signal instead of individual peaks. Stationary phase degradation can not be the source of peaks. Even a defect at a single point in the stationary phase does not result in peaks. Peaks are due to individual compounds originating from a contamination source somewhere in the GC system.

4.6.2 Measuring Column Bleed

Column bleed can be difficult to accurately measure. There are many contributors to the background signal with column bleed being only one of them. There are several methods used to measure column bleed. The detector signal at the column's temperature limit can be used. This signal level is obtained from the data system/integrator or detector output signal on the GC. Other sources contribute to the magnitude of this signal; thus the contribution from the column is difficult to determine. A better technique is to compare the signal at the upper temperature limit to the one obtained at a lower temperature (usually between 50-100°C). Most of the contributions from other sources are constant at the two temperatures; thus the difference in the signals is primarily due to the amount of column bleed. This technique is used by most column manufacturers in their quality control tests.

Collecting a bleed profile is another method to measure column bleed. If the same GC and data system settings are used, bleed profiles can be directly compared. While this method does not provide absolute numbers, column bleed changes or differences are readily visible upon comparison.

4.6.3 Sensitivity Considerations

The size of the baseline rise as drawn on integrator paper or data system screen can be deceptive. The actual size of the baseline rise is dependent on the attenuation or scale setting for the recording device. For low level analyses, the chromatograms are expanded (small attenuation values) so that the small peaks are visible. At these settings, any rise in the baseline appears to be quite large. When the chromatograms are reduced in scale, the rise appears much smaller or is not visible at all. The amount of column bleed is the same, but the rise is amplified in the former situation.

Column bleed is a greater concern for low level analyses. The very small baseline rise present at lower temperatures becomes more visible and has a greater impact on peak integration. The large baseline rise at high temperatures goes off scale and any peaks in this region are not visible. The chromatogram has to be redrawn at smaller scale to keep the baseline on scale; however, the peaks of interest may become too small to be visible. Column bleed becomes less of a concern at higher sample concentrations. Even large increases in column bleed are not readily visible and do not interfere with the peaks.

4.6.4 Detector Considerations

Due to the extreme sensitivity of some detectors for some stationary phase degradation products, some stationary phases appear to have substantially greater bleed than others. For example, the baseline rise at higher temperatures for a cyanopropyl substituted polysiloxane is much greater with a nitrogen phosphorus detector (NPD) than with a flame ionization detector (FID). The NPD is more sensitive to the presence of nitrogen than a FID. The amount of column bleed is the same for the two detectors, but the NPD is more sensitive to the bleed species and produces a much larger baseline rise at the same recorder settings. If the larger baseline rise does not create any problems, there is no damage to the column or GC.

From a chromatographic perspective, mass spectrometers are not significantly different from other, less complex GC detectors. A mass spectrum of the baseline contains any of the stationary phase degradation products eluting from the column when the spectrum was collected. The spectrum can be obtained anytime during the chromatographic run, but the bleed is most intense at the upper temperature limit of the column. A mass spectrum of the baseline is not the spectrum of a single column bleed compound, but of the sum of the stationary phase degradation products. The major fragments in the mass spectrum usually correspond to the more stable degradation products and are often the same as the mass fragments of the larger, less stable degradation products. The larger mass species are usually only visible for higher bleed columns. The mass spectra of the most common stationary phases can be found in Appendix D.

A common error is to attribute a rising baseline to excessive column bleed because siloxane materials are identified by the mass spectrometer's library program. Some types of contaminants are often misidentified as siloxanes. Silicone oils used in flow controllers and pressure regulators, and silicone septum bleed species often produce mass spectra very similar to those of many common stationary phases.

Adding to the problem is the presence of the major stationary phase degradation products at all times. In cases where the stationary phase bleed ions are the most abundant (e.g., higher column temperatures, older column, low compound concentrations), the mass spectrum is reasonably close enough to authentic column bleed that the library search labels a peak or baseline areas as a polysiloxane. From this match, a conclusion of excessive column bleed is made. Obtaining reference mass spectra for a new column in a verified clean GC/MS system is recommended to minimize the possibility of misidentifying column bleed.

4.6.5 Minimizing Column Bleed

Complete elimination of column bleed is not possible. Column bleed continues to increase with column use. The increase in bleed is very small for a new column then it starts to elevate at a much faster rate. Eventually, the bleed increases to a level that renders the column unusable. The onset of excessively high bleed is dependent on the stationary phase polarity, temperature of column usage and the oxygen concentration in the carrier gas.

There are two conditions that increase or accelerate column bleed - high temperature and oxygen. Stationary phase degradation increases with temperature. Using columns at least 20-30°C below their upper temperature limit avoids the temperature region where column bleed is the highest. Oxygen increases the rate of stationary phase degradation. Maintaining a leak free system and using high quality (low oxygen concentration) carrier gas slows down the onset of higher bleed.

4.7 Selecting Stationary Phases

Selecting the best stationary phase is a mixture of science, experience and a bit of guess work. A variety of factors need to be considered before deciding which stationary phase is best. The first step is to consult the literature. It is very likely that chromatograms or analysis information similar to the analysis of interest can be found in a published paper. Stationary phase selection is more certain and easier even with partial information from previous work. GC and column manufacturers are also good sources of applications information. Chromatograms are usually employed to help market and sell their products, and many manufacturers offer technical support services at no cost.

Selecting a stationary phase without any previous information is an inexact science. It is usually easier to determine which stationary phases will not work. This narrows the choices to the ones most likely to work. Before selecting a stationary phase, details about the compounds and their concentrations, available GC hardware, analysis time considerations, and any other important analysis factors need to be determined. These may influence stationary phase (and column size) selection.

It is fairly common for more than one stationary phase to separate all of the sample compounds. There are plenty of other factors that may influence stationary phase selection. When selecting a stationary phase, the following factors should be considered:

1) *Previous separation information* - Examine the compound separations on other stationary phases. These separations are often very helpful in selecting another stationary phase. Literature references and application notes are great sources of this type of information. If other columns are available in the laboratory, try the analysis on these columns providing the column dimensions are suitable. One of these columns may provide satisfactory separation.

2) *Selectivity* - Determine the potential compound/stationary phase interactions. If the compounds have different dipoles or can hydrogen bond, consider a selective stationary phase with these interactions. If the compounds are hydrocarbons or are part of a homologous series, a selective stationary phase is probably not necessary.

3) *Polarity* - Use the most non-polar stationary phase that provides the required separations. Non-polar stationary phases are more versatile, have longer lifetimes and generally exhibit better efficiency.

4) *Temperature limits* - Compounds with high boiling points or molecular weights require high column temperatures to avoid extremely long retention times. The lower temperature limits of polar stationary phases restricts their use to low and moderate boiling point compounds. When multiple columns are in the same GC oven, a lower temperature limit of one column may restrict the temperature range of the other.

5) *Compound activity* - Columns with non-polar stationary phases usually are the most inert. Greater peak tailing or adsorption may be experienced with polar stationary phases.

6) *Analysis time* - Some stationary phases provide satisfactory separations in less time.

7) *Capacity* - Stationary phases similar in polarity to the compounds have greater capacity for those compounds. If one or more of the compounds are present at very high levels, the overloaded peak may interfere with some of the other peaks.

8) *Bleed* - In general, non-polar stationary phases have lower bleed. Compounds with high boiling points or molecular weights elute in the higher temperature regions where column bleed is more severe. Non-

polar stationary phases not only bleed less, but the onset of higher bleed occurs at higher temperatures.

9) *Selective detectors* - Avoid stationary phases containing the species or functional group that generates a strong response with a selective detector. Extreme baseline rises and higher noise levels usually occur.

10) *Versatility* - For multiple analyses, different stationary phases may be required for optimal separation. In some cases, several analyses can be performed with a single stationary phase providing some sacrifices are acceptable. This reduces the number of columns needed which may reduce complexity and cost.

11) *Critical separations* - The separation of some peaks may be more important than others. Greater separation of these peaks with less separation of others may be desired. This helps to compensate for errors, concentration induced retention shifts and variable matrix interferences that can reduce the separation of the critical peaks.

5   CAPILLARY GC COLUMNS: DIMENSIONS

5.1 Introduction

Capillary columns are available in a wide variety of sizes. Combined with the different stationary phases, there are hundreds of different possible columns. The large number of column possibilities seems to make column selection difficult. A knowledge of the effect of each column dimension on the chromatography greatly facilitates column selection and use. Column length and diameter mainly affect peak resolution while film thickness mainly affects peak retention. The stationary phase is responsible for the separation of the peaks. If better separation is needed, changing a column dimension rarely improves the situation; changing the stationary phase or column temperature is required. If better resolution is needed, changing one or more of the column dimension is required.

5.2 Column Length

The most frequently used capillary column lengths are between 12 and 30 meters. Columns as long as 150 meters and short as 5 meters are commercially available. Differences of 10-30 meters between available lengths are common. Fortunately, column lengths are easily shortened by cutting the tubing and lengthened by joining two pieces with an union. Most column manufacturers will supply columns in lengths other than their regular offerings on a special request basis. Using the standard column lengths is easier and makes the most sense (and usually less expensive) since column length is not a critical column dimension in most cases.

5.2.1 Length and Efficiency/Resolution

The number of theoretical plates is directly proportional to column length (Table 5-1). For example, doubling column length doubles the total number of theoretical plates or halving the length reduces the number of plates by half. Increasing the number of theoretical plates results in better resolution, but the resolution increase is not directly proportional to the increase in the number of theoretical plates.

Resolution is proportional to the square root of the number of theoretical plates. Since the number of theoretical plates and column length are directly proportional, resolution is proportional to the square root of column length (Table 5-1). For example, doubling column length increases resolution by 41% (not 100%) or halving the length reduces resolution by 29% (not 50%). In practice, resolution changes are smaller than expected. Usually resolution increases by 25-35% upon doubling the column length and resolution decreases by 15-25% upon halving the column length.

The square root relationship means column length has to increase by four times to double the resolution by changing column length alone. For shorter columns, this is not much of a problem. A 15 meter column can be replaced with one 60 meters long since 60 meter columns are readily available. For longer columns, this is a problem. A 30 meter column has to be increased to 120 meters to double the resolution; 60 meters is the longest standard length columns available.

Table 5-1:   Changes in Efficiency and Resolution with Length

Length (m)	$n^{(1)}$	$R^{(2)}$
5	23,765	0.41
12	57,035	0.63
15	71,295	0.71
25	118,825	0.91
30	142,590	1.00
50	237,650	1.29
60	285,180	1.41
75	356,475	1.58

1.   Theoretical maximum efficiency for 0.25 mm I.D. column; k = 5.
2.   Resolution normalized to the 30 meter column

The square root relationship of length and resolution is beneficial when shortening column lengths. A 30 meter column has to decrease to 7.5 meters before the resolution is halved. Substantial length of column can be removed without negatively affecting resolution. For example, removing 1 meter from a 30 meter column reduces resolution by less than 2%. Trimming short pieces off a column such as during column installation has virtually no impact on resolution.

Changes in column length have less of an influence on resolution as column lengths become longer. Adding 5 meters to a 60 meter column results in a 4% increase in resolution; however, adding 5 meters to a 10 meter column increases resolution by 22%. Removing 5 meters from a 60 meter column results in a 4% decrease in resolution; however, removing 5 meters from a 10 meter column decreases resolution by 29%.

5.2.2 Length and Retention

Retention as measured by the partition ratio (k) is not affected by a change in column length; however, retention time is affected by a change in the column length. For isothermal temperature conditions, the change in the retention time is roughly proportional to the change in column length. For example, doubling column length almost doubles the retention times or halving the length reduces retention time by almost half (Table 5-2). For temperature program conditions, the change in retention time is less than for isothermal conditions. In most cases, the change in retention time is between 25 and 75% (Table 5-2).

5.2.3 Length and Pressure

The column head pressure is directly proportional to column length. For example, doubling the column length doubles the head pressure required to maintain a particular carrier gas average linear velocity (or flow rate) and halving the column length reduces the required head pressure by half.

Most GC system work better when the head pressure is greater than 1 psig. Controlling and consistently setting the average linear velocity is very difficult when the pressure gauge reads below 1 psig. Megabore (0.53 mm I.D.) columns require low head pressures to deliver the proper average linear velocity. Even with helium as the carrier gas, 15 meter Megabore columns barely have enough back pressure to generate a head pressure greater than 1 psig.

Table 5-2:   Changes in Retention with Length

Temperature	Length (m)	Retention Times (min)		
		Peak 1	Peak 2	Peak 3
130°C isothermal	15	1.81	2.81	5.39
	30	3.32	5.14	10.67
	60	5.39	9.82	20.41
125°C for 1 min	15	1.89	2.73	4.14
125-225°C at 10°/min	30	3.16	4.25	5.92
	60	5.69	7.07	8.99
75°C for 1 min	15	4.97	6.71	8.85
75-225°C at 10°/min	30	6.47	8.31	10.51
	60	9.86	11.85	14.18

Peak 1:   1-chloro-4-nitrobenzene   *Peak 2:*   2,4-dichloronitrobenzene
Peak 3:   2,4-dinitrochlorobenzene   *Column:*   DB-1, 0.32 mm I.D., 0.25 µm
Carrier: Helium at 35 cm/sec at initial temperature

Narrow diameter columns have the opposite problem of those with wide diameters. Long lengths of narrow diameter columns require high head pressures. For example, a 60 m x 0. 25 mm I.D. column at 100°C requires about 34 psig (235 kPa) of helium to deliver an average linear velocity of 30 cm/sec. The pressure gauges on many GC's have an upper range of 30 psig (210 kPa). The pressure gauge has to be changed to a higher range one to accommodate the long, narrow column.

Mass spectrometers are under vacuum and the column exit is in this vacuum. The vacuum helps to pull the carrier gas through the column. Significantly lower head pressures are required to maintain suitable carrier gas velocities in GC/MS systems. Even 15 m x 0.25 mm I.D. columns using helium as a carrier gas operate around 1 psig. Usually longer columns are necessary with GC/MS systems especially if wider diameter columns are being used.

5.2.4 Length and Bleed

Column bleed increases as column length increases. The bleed increase is usually small. In theory, the bleed change should be roughly proportional to the change in the column length. In practice, the column-to-column bleed variation between columns of the same description is close to the difference in bleed between columns of different lengths. Rarely is the increased column bleed with longer columns ever a factor when selecting column length.

5.2.5 Length and Cost

For a particular stationary phase and diameter, column cost is primarily dependent on column length. Cost and column length are almost proportional. Using excessively long columns can be a costly error. Since resolution is dependent on the square root of column length, increasing resolution by increasing column length is an expensive method. Along with the additional column cost, the longer run times is not a cost effective situation in many laboratories.

The cost of two 30 meters columns is 10-20% greater than one 60 meter column of the same description. Two 30 meter columns require twice the packaging, handling and testing than one 60 meter column. It is tempting to cut a longer column into shorter lengths to save some money. In most cases, there are no significant differences between the individual lengths from a single column. Differences between the individual pieces start to appear with increases in stationary phase polarity, length and stationary phase film thickness, and a decrease in diameter of the original column. The possibility of one of the pieces being different increases as the number of pieces cut from the original column goes up.

5.2.6 Selecting Column Length

Most analyses can be easily done on 12-30 meter columns. Columns longer than 30 meters should only be used when all other reasonable methods of improving resolution have been exhausted or are unavailable. Usually, the increase in analysis time and cost are greater than the return in resolution improvement. Only very complex samples or the most difficult to resolve compounds should be analyzed using a 60 meter column. Many analyses are performed with columns that are too long. Unfortunately, columns are available in fairly large increments of length. One length is too short while the next longer length is too long. Often the longer column has to be used with its excessive resolution and run times. Increasing the column temperature often eliminates some of the excess analysis time while still obtaining an increase in resolution. Occasionally, a longer column may be needed to increase carrier gas head pressures in GC/MS systems. Sometimes, reducing the column diameter instead of increasing column length is better method to increase the head pressure.

5.3 Column Diameter

The most frequently used capillary column inner diameters are between 0.25 and 0.53 mm. Columns with diameters as small as 0.05 mm are available, but they are rarely used for GC applications. The common name of Megabore is often applied to 0.53 mm I.D. columns. The differences between the available diameters seem small, but the relative differences are quite large (30-70%).

5.3.1 Diameter and Efficiency/Resolution

The number of theoretical plates is inversely proportional to column diameter. In other words, theoretical plates per meter increases as column diameter decreases (Table 5-3). Decreasing column diameter is one of the easiest methods to increase efficiency. It appears that very small diameter columns are the best ones to use for maximum efficiency. While this is true, very small diameter columns are not practical in the standard laboratory environment. The smallest diameter column easily used for most GC applications is about 0.18 mm I.D.. Increasing the number of theoretical plates results in better resolution, but the resolution increase is not directly proportional to the increase in the number of theoretical plates.

Resolution is proportional to the square root of the number of theoretical plates. Since the number of theoretical plates and column diameter are directly proportional, resolution is proportional to the square root of column diameter (Table 5-3). For example, halving column diameter increases resolution by 41% (not 100%) or doubling the diameter reduces resolution by 29% (not 50%). In practice, resolution changes are smaller than expected. Usually resolution increases by 25-35% upon halving the column diameter and resolution decreases by 15-25% upon doubling the column diameter.

The square root relationship means that column diameter has to decrease by four times to double the resolution by changing column diameter alone. For larger diameter columns, this is not much of a problem. Changing a 0.53 mm I.D. column to a 0.25 mm I.D. column

Table 5-3:   Changes in Efficiency and Resolution with Diameter

Diameter (mm)	$n^{(1)}$	$R^{(2)}$
0.10	375,000	1.62
0.18	198,025	1.18
0.20	178,220	1.12
0.25	142,575	1.00
0.32	111,390	0.88
0.45	79,210	0.75
0.53	67,255	0.69

1.   *Theoretical maximum efficiency for 30 meter column; k = 5.*
2.   *Resolution normalized to the 0.25 mm I.D. column.*

increases the number of theoretical plates by over two fold with a resolution increase of approximately 45%. Attempting the same improvement with a smaller diameter column is difficult because very narrow diameter columns are required to obtain the same results. A 0.25 mm I.D. column has to be replaced with a 0.12 mm I.D. column to obtain similar improvements. Columns of this small diameter are difficult to use and are not recommended for routine GC applications.

5.3.2 Diameter and Retention

For isothermal conditions, the partition ratio is inversely proportional to column diameter. Retention times are similarity affected, but not in the same linear manner as k. For example, halving column diameter doubles retention or doubling the diameter reduces retention by half, but the retention time changes are not quite directly proportional (Table 5-4). For temperature program conditions, the change in retention is less than for isothermal conditions. In most cases, the amount of retention change is between 25 and 75% (Table 5-4). Rarely is the difference in retention between columns of the various diameters ever a factor when selecting column diameter.

5.3.3 Diameter and Pressure

The column head pressure varies with the diameter roughly according to an inverse square relationship (Table 5-5). For example, reducing column

Table 5-4: Changes in Retention with Diameter

Temperature	Diameter (mm)	Retention Times (min)		
		Peak 1	Peak 2	Peak 3
130°C isothermal	0.25	4.05	6.47	12.71
	0.32	3.32	5.14	10.67
125°C for 1 min	0.25	3.71	4.95	6.75
125-225°C at 10°/min	0.32	3.16	4.25	5.92
75°C for 1 min	0.25	7.61	9.52	11.77
75-225°C at 10°/min	0.32	6.47	8.31	10.51

Peak 1: 1-chloro-4-nitrobenzene   Peak 2: 2,4-dichloronitrobenzene
Peak 3: 2,4-dinitrochlorobenzene   Column: DB-1, 30 meters, 0.25 µm
Carrier: Helium at 35 cm/sec at initial temperature

Table 5-5: Changes in Column Head Pressure with Diameter

Diameter (mm)	Pressure (psig)		
	H_2	He	N_2
0.05	404	497	222
0.10	93.3	114	49.7
0.18	25.9	32.3	14.0
0.20	20.7	25.7	11.3
0.25	12.9	16.0	7.1
0.32	7.8	9.5	4.3
0.45	3.9	4.8	2.2
0.53	2.8	3.4	1.6
u (cm/sec)	50	30	15

Column length: 30 meters   Column temperature: 100°C

diameter by half results in a head pressure increase of over 4 times to maintain a specific carrier gas average linear velocity. For this reason, smaller diameter columns are usually offered in shorter lengths to compensate for the higher head pressure requirements.

The standard pressure gauges on many GC's have an upper range of 30 psig (210 kPa). The pressure gauge may have to be changed to one with a higher range to accommodate long, narrow columns. When using helium as the carrier gas, changing to hydrogen reduces head pressures by 20-25%. Using hydrogen makes the use of narrow diameter columns easier. Reducing column head pressures has an additional benefit in that higher pressure require more frequent septum changes and increases the probability of carrier gas leaks.

5.3.4 Diameter and Bleed

Column bleed increases as column diameter increases. The increased surface area of the larger diameter column requires a greater mass of stationary phase to maintain the same film thickness. The greater mass of stationary phase results in a greater amount of degradation products eluting from the column. In practice, the bleed increase with wider diameter column is usually small. The column-to-column variations in bleed for a given diameter are usually greater than the differences between the bleed levels of the different diameters. The increased bleed with larger diameter columns is rarely a factor when selecting column diameters.

5.3.5 Diameter and Capacity

Column capacity increases as column diameter increases in a fairly proportional manner (Table 5-6). The increased surface area of the larger diameter column contains a greater mass of stationary phase to maintain the same film thickness. Since there is more stationary phase in a given length of column, a greater amount of a compound can be dissolved in the stationary phase before overloading occurs. A list of capacities for a variety of column diameters and film thicknesses can be found in Appendix E.

Column capacity is also dependent on the polarity of the stationary phase and the compounds. Close polarities of compounds and stationary phases result in higher capacities. For example, a polar stationary phase has a greater capacity for a polar compound than a non-polar one. This makes it possible for a smaller diameter, polar stationary phase column to have higher capacity for a polar compound than a larger diameter, non-polar stationary phase column.

Table 5-6:   Capacity and Diameter

Diameter (mm)	Capacity Range (ng)*
0.18-0.20	25-50
0.25	50-100
0.32	50-125
0.53	100-250

Film thickness: 0.25 μm
* Similar polarity stationary phase and compound

5.3.6 Diameter and Carrier Gas Volume

Smaller diameter columns use a lower volume of carrier gas per unit time (mL/min) to operate at their most efficient level. This may not seem significant, but there are cases were carrier gas volume is important. Many bench top GC/MS systems have a maximum pumping capacity of 1-2 mL/min of carrier gas. The carrier gas volumes easily exceed this range for columns with inner diameters of 0.32 mm or greater (Table 5-7). Decreased sensitivity occurs at higher carrier gas volumes. For portable GC systems or in locations where high purity gases are difficult or expensive to obtain, the low carrier gas consumption of smaller diameter columns is beneficial.

Table 5-7:   Flow Rate and Diameter

Diameter (mm)	Flow Rate at Column Exit (mL/min)		
	H_2	He	N_2
0.18	1.65	1.19	0.34
0.20	1.72	1.21	0.37
0.25	2.07	1.39	0.49
0.32	2.76	1.78	0.71
0.45	4.61	2.88	1.28
0.53	6.08	3.76	1.72
u (cm/sec)	50	30	15

Column Temperature: 100°C

5.3.7 Diameter and Injector Efficiency

Directly stated, most capillary injectors work better at high carrier gas flow rates. The higher volume of carrier gas sweeps the vaporized sample into the column much faster. Larger diameter columns have higher flow rates at the same linear velocity as a smaller diameter column. In cases where the sample is not significantly trapped or focussed at the front of the column, the faster transfer rate results in a shorter sample band. Better efficiency is obtained when the sample band is shorter.

For splitless injections, better efficiency for low k peaks is obtained with a 0.32 mm I.D. column than a 0.25 mm I.D.. The faster transfer of the sample and subsequent short band width overcomes the loss of efficiency of the larger diameter column. The gain in efficiency decreases as retention increases until better efficiency is obtained with the 0.25 mm I.D. column.

Split injectors rapidly transfer the sample into the column, and splitless and on-column injectors utilize a focusing effect to maintain reasonable injector efficiencies. Megabore direct injectors usually do not utilize any type of focusing effect, thus high carrier gas flow rate are needed to rapidly transfer the sample into the column. For this reason, only 0.45-0.53 mm I.D. columns are suitable for use in Megabore direct injectors. Smaller diameter columns require much lower carrier gas flow rates, thus slow sample transfer occurs with the resulting poor injector efficiency.

5.3.8 Diameter and Breakage

The largest internal diameter fused silica column regularly available is 0.53 mm. Larger diameter tubing proves to be too fragile and excessive breakage occurs in normal use. Smaller diameter columns are stronger and break less often. It is unusual for 0.53 mm I.D. tubing to be available on a cage that is less than 7 inches (18 cm) in diameter. A smaller diameter cage places to much stress on the large bore tubing and excessive breakage occurs. Tubing with diameters of 0.25 mm I.D. or less has been wound in diameters of 1-2 inches without any breakage problems.

5.3.9 Diameter and Cost

As a rule, tubing costs increase as diameter increases. For this reason, larger diameter columns are slightly higher in price than a smaller diameter column of the same length. The price for 0.05 mm and 0.10 mm I.D. columns increases due to the higher tubing cost and manufacturing difficulties. Columns with these diameters are normally not used for GC work, thus their higher price should not be an issue.

5.3.10 Selecting Column Diameter

The most common column diameters for use with split and splitless injectors are 0.25 and 0.32 mm. Column diameters of 0.25 mm are used when higher efficiency and lower carrier flow rates are needed. Smaller diameter columns are best suited for complex samples, difficult to resolve compounds and GC/MS systems. The higher efficiency of the 0.18 and 0.20 mm I.D. columns may be beneficial; however, these small diameter columns are often only available in shorter lengths to reduce carrier gas head pressures. The shorter lengths may negate the efficiency improvements with the smaller diameter column. Columns diameters of 0.32 mm are used when higher capacity and high carrier gas flow rates are needed. These column diameters are often used with on-column and splitless injectors. The loss in efficiency is often compensated by the gains in capacity and injector efficiency. Columns with 0.53 mm I.D. diameters are intended for use with Megabore direct injectors. Nothing prevents 0.53 mm I.D. columns from being used in split, splitless or on-column injectors, but the loss in efficiency is rarely worth any of the potential benefits. The high capacity and carrier gas flow rate of 0.53 mm I.D. columns are probably the only valid reasons to use such a large diameter column without a Megabore direct injector.

5.4 Column Film Thickness

The most frequently used film thicknesses are between 0.25 and 1.0 µm. Columns with films as thin as 0.1 µm and thick as 10 µm are commercially available. Film thicknesses outside the range of 0.1 to 5 µm present a large series of different problems which limits their functionality. The differences between the available film thicknesses are fairly large (200-400%). Film thickness is not manipulated or an option with most PLOT columns.

5.4.1 Film Thickness and Retention

For isothermal conditions, retention (k) is directly proportional to film thickness. Retention times are similarly affected, but not in the same linear manner as k. For example, doubling the film thickness doubles retention or halving the film thickness reduces retention by half, but the retention time changes are not quite directly proportional (Table 5-8). For temperature program conditions, the change in retention is less than for isothermal conditions. In most cases, the amount of retention change is between 25 and 75% (Table 5-8).

One question that arises is why use film thickness to change retention when adjusting column temperature is easier and more flexible? Some analyses require temperature conditions where the GC or column is at the extreme of its temperature range. Adjusting retention using column temperature may not be practical or possible in these cases. Most GC's are limited to a lower temperature of about 35°C without the use of cryogenic cooling. Difficulties arise when greater retention is desired to improve peak separation or to move a peak away from the solvent front, but the GC is already at its lowest temperature. Changing to a thicker film column increases retention without the need to lower the column temperature. In most cases, a slightly higher column temperature than originally used with the thinner film column can be used. A higher initial temperature in a

Table 5-8:   Changes in Retention with Film Thickness

Temperature	Film (µm)	Retention Time (min)		
		Peak 1	Peak 2	Peak 3
150°C isothermal	0.10	1.92	2.30	3.21
	0.25	2.42	3.27	5.33
	1.0	6.22	10.09	19.29
125°C for 1 min	0.10	2.35	3.02	4.26
125-225°C at 10°/min	0.25	3.16	4.26	5.93
	1.0	6.52	8.29	10.49
75°C for 1 min	0.10	5.05	6.71	8.81
75-225°C at 10°/min	0.25	6.77	8.63	10.85
	1.0	11.30	13.41	15.80

Peak 1:   1-chloro-4-nitrobenzene   Peak 2:   2,4-dichloronitrobenzene
Peak 3:   2,4-dinitrochlorobenzene   Column: DB-1, 30 m x 0.32 mm I.D.
Carrier: Helium at 35 cm/sec at initial temperature

temperature program reduces the time required to cool down the GC after the program has finished. This decreases the time between sample injections, thus increasing the number of samples that can be run per day or work shift.

With few exceptions, most columns have upper temperature limits of 220-350°C. Difficulties arise when less retention is desired, but the column is already at its upper temperature limit. Changing to a thinner film column decreases retention without the need to increase the column temperature. In some cases, a lower temperature than originally used with the thicker film column can be used. The lower column temperature usually increases column lifetime and reduces bleed and may benefit compounds that decompose in the column at high temperatures. The lower column temperature does not subject the compound to the higher temperatures where greater degradation may occur.

Column temperature and retention are not directly proportional - they are related by an inverse natural log (ln) function. This means that when column film thickness is changed, a much smaller relative temperature change is needed to maintain the same retention as before. For isothermal conditions, the change in column temperature is related to the inverse natural log of the change in the film thickness. For example, if the compound peak has a retention time of 10 minutes at 100°C with a 0.25 µm film column, 138°C is needed for a 1 µm film column to keep the peak at 10 minutes. For temperature program conditions, it is more difficult to predict the magnitude of the retention changes. The inverse natural log relationship between retention and column temperature still applies, but changes in the initial and final temperatures, the ramp rate and the hold times all affect the retention shift. The initial temperature and its hold time has the greatest impact on the earlier eluting peaks; the final temperature and ramp rate has the greatest impact on the later eluting peaks. Peaks eluting in the middle part of the chromatogram are influenced to different degrees by all of the program conditions.

5.4.2 Film Thickness and Efficiency/Resolution

When properly calculated (i.e., using a peak with k > 5), the number of theoretical plates decreases as film thickness increases (Table 5-9). When incorrectly calculated (i.e., using a peak with k < 5), the number of theoretical plates increase as film thickness increases (as long as the peaks remain at

Table 5-9:   Changes in Efficiency with Film Thickness

Film Thickness (µm)	n
0.25	51,700
1.0	48,400
3.0	30,600
5.0	20,000

Column: 15 m x 0.32 mm I.D.   *Peak at k = 7*
Carrier: Helium at 35 cm/sec

k < 5 on the thicker film column) (Table 5-9). This behavior has a direct impact on using film thickness to improve peak resolution.

One method to improve resolution is to increase the retention of the peaks. This can be accomplished by lowering the column temperature or increasing the film thickness. When using film thickness to improve resolution via an increase in retention, there is a guideline to follow. If the peaks have k values less than 5, increasing the column film thickness increases the resolution of peaks providing the peaks still have k values less than 5 on the thicker film column. If the peaks have k values greater than 5, increasing the column film thickness does not improve or decreases the peak resolution.

The improved resolution of peaks with k's less than 5 upon increasing film thickness is more of a guideline than a rule. The amount of resolution gain or loss is influenced by the stationary phase, column temperature and the compounds. The biggest problem with using film thickness to improve resolution is when the peaks in the chromatogram elute over a wide k range. Improving resolution of the low k peaks by increasing the film thickness often reduces the resolution of the high k peaks. When this happens, increasing the ramp rate or programming to a higher final temperature sometimes reduces the extent of resolution loss of the high k peaks.

5.4.3 Film Thickness and Capacity

Column capacity increases as film thickness increases in a fairly proportional manner (Table 5-10). A greater amount of compound can be dissolved in the greater mass of stationary phase in a given length of column. When attempting to increase capacity, some of the increased retention (and possible resolution change) with the thicker film column can be overcome if the column temperature is raised. A list of capacities for a variety of column diameters and film thicknesses can be found in Appendix E.

Column capacity is also dependent on the polarity of the stationary phase and the compounds. Close polarities of compounds and stationary phases result in higher capacities. For example, a polar stationary phase has better capacity for a polar compound than for a non-polar one. This makes it possible for a thinner film, polar stationary phase column to have higher capacity for a polar compound than a thicker film, non-polar stationary phase column.

Table 5-10: Capacity and Film Thickness

Film Thickness (µm)	Capacity Range (ng)*
0.10-0.20	20-50
0.25	50-125
0.50	125-250
1.0	250-500
3.0	500-1000
5.0	1000-2000

Column Diameter: 0.32 mm
* *Similar polarity stationary phase and compound*

5.4.4 Film Thickness and Bleed

Column bleed increases as column film thickness increases. The greater mass of stationary phase results in a greater amount of degradation products eluting from the column. Very thick film columns (≥ 3 µm) usually have a lower value for the upper temperature limit than a thinner film counterpart. The high bleed of these very thick film columns requires their upper temperature limits to be set at lower values. Non-polar stationary phases have lower bleed levels than more polar stationary phases. This makes it possible for a non-polar, thicker film column to have lower bleed than a polar, thinner film column. Due to the higher bleed of polar stationary phases, film thicknesses of 3 µm or greater are usually available only for non-polar stationary phases.

5.4.5 Film Thickness and Inertness

The surface of the tubing is the site of activity in capillary columns. Interaction of the compounds with these sites is undesirable and needs to be avoided. The interaction results in reversible (peak tailing) or irreversible (loss of peak size) adsorption. Thicker films shield the compounds from these active sites and reduces the number of interactions. Peaks that tail on one column may show reduced adsorption on a corresponding thicker film column. Problems are encountered when the film thickness is reduced and activity problems occur with the thinner film column that were not experienced with the thicker film column. When attempting to improve inertness, some of the increased retention (and possible resolution change) with the thicker film column can be overcome if the column temperature is raised.

5.4.6 Selecting Column Film Thickness

Improving the resolution for low k compound peaks is the primary reason for using a thick film column. Reducing retention of highly retained compound peaks when limited by column temperature is the primary reasons for a thin film column. For most analyses, a standard film thickness is recommended. For 0.20-32 mm I.D. columns, the standard film thickness is 0.2-0.5 µm; for 0.53 mm I.D. columns, it is 1.0-1.5 µm. Improving capacity or inertness by increasing the film thickness is accompanied by increased retention and bleed. Some of the increase in retention and possible change in resolution can be overcome by using a higher column temperature; however, the higher bleed of the thicker film column at the higher temperature can be a problem.

5.5 Manipulating Multiple Column Dimensions

Changing any of the column dimensions often affects more than one performance parameter. Also, the same parameters are often affected by each column dimension. Sometimes, these interconnected relationships can be advantageous. An example best illustrates this concept. A 15 m x 0.25 mm I.D. column is being used, but a lower carrier gas flow (volume) into the detector is desired. This is best accomplished by reducing the column diameter. Decreasing column diameter also increases efficiency and retention. While the increase in efficiency is good, the increase in retention may not be so good. Realizing that retention and efficiency decrease with shorter columns, using a shorter column may be a satisfactory adjustment.

A 12 m x 0.2 mm I.D. column is available - does this column provide all of the desired improvements without too much of a sacrifice? Assuming a worst case scenario (i.e., isothermal), the 0.2 mm I.D. column is 25% more retentive than the 0.25 mm I.D. column; the 12 meter column is 20% less retentive than the 15 meter column. Concerning retention, these two dimension changes nearly cancel each other. An average 0.2 mm I.D. column has n/m = 5350 and an average 0.25 mm I.D. column has n/m = 4275. The total number of theoretical plates for the 12 m x 0.2 mm I.D. column is 64,200 and 64,125 for the 15 m x 0.25 mm I.D. column. Again, the two dimension changes nearly cancel each other. Using a 12 m x 0.2 mm I.D. instead of a 15 m x 0.25 mm I.D. column provides the lower carrier gas flow into the detector while causing relatively insignificant changes in retention and efficiency. There is a higher carrier gas head pressure and lower capacity affect when changing to a smaller diameter column. As long as this is not a problem, the 12 m x 0.2 mm I.D. column can be successfully used to lower carrier gas flow.

It is important to realize that changing a column dimension changes other performance parameters. Sometimes these parameters changes are not acceptable. It is equally important to realize is another column dimension can be altered to counteract some of the changes to the parameters. Sometimes the parameter changes can not be offset by changing another dimension or another change occurs that is not acceptable. All of the various performance parameters and how they are related to column dimensions have to be reconciled. Table 5-11 summarizes the major relationships between column behavior and column dimensions.

Table 5-11:   Relationship Between Column Behavior and Dimensions

Increase[1]	Length	Diameter	Film Thickness
Efficiency	Increase	Decrease	Decrease[2]
Resolution	Increase	Decrease	Increase (k<5) Decrease (k$\geq$5)
Retention time	Increase	Decrease	Increase
Bleed	Increase	Increase	Increase
Capacity	No affect	Increase	Increase
Head Pressure	Increase	Decrease	No affect
Inertness	No affect	No affect	Increase
Carrier Gas Flow	No affect	Increase	No affect
Injector Efficiency	No affect	Increase	No affect

1. *Note: To increase each parameter not improve*
2. *For peaks with a k value greater than 5*

6 CARRIER GAS

6.1 Carrier Gas and Capillary Columns

The influence of the carrier gas in capillary column chromatography is often underappreciated and neglected. The type of carrier gas and its velocity through the column have a significant impact on resolution and retention time. The proper selection of the type and velocity of the carrier gas is critical for full exploitation of the high efficiencies of capillary columns. Small variations or deviations in the carrier gas velocity may result in noticeable losses in efficiency.

6.2 Linear Velocity versus Flow Rate

The linear velocity is the speed (cm/sec) of the carrier gas as it travels through the column. The flow rate is the volume of carrier gas passing through the column per unit time (mL/min). The linear velocity is more important and meaningful for capillary columns. The linear velocity affects efficiency and retention time. There are situations where the carrier gas flow rate is important; however, they involve injectors and detectors and do not have a direct impact on column efficiency.

Carrier gas linear velocity is not uniform throughout the column, hence the average linear velocity ($\bar{u}$) is used for measurement purposes. There is an optimal linear velocity ($\bar{u}_{opt}$) for each type of carrier gas where maximum column efficiency is obtained. There are small differences between the optimal average linear velocities for columns of different diameters, film thickness and length. Because they are usually small, these differences are often ignored without any negative consequences.

6.3 Van Deemter Curves

The relationship between average linear velocity and efficiency is described by the Van Deemter equation (Equation 6-1). If efficiency as h_{min}

is plotted against the average linear velocity for a given column and set of conditions, a Van Deemter curve is obtained (Figure 6-1).

The optimal linear velocity ($\bar{u}_{opt}$) is where the curve is at its lowest point which corresponds to the smallest value of h_{min} (i.e., point of greatest efficiency). Also important, Van Deemter curves show that using too high (right side of the curve) or low (left side of the curve) a linear velocity results in a loss of efficiency. However, this is not always a completely undesirable situation.

Using an average linear velocity greater than $\bar{u}_{opt}$ is often recommended. This value is called the optimal practical gas velocity (OPGV) and is 1.25-2 times $\bar{u}_{opt}$. The OPGV is where the maximum efficiency per unit time is obtained. The average liner velocity corresponding to $\bar{u}_{opt}$ provides the highest efficiency, but average linear velocities in the OPGV range sacrifice a small amount of efficiency for substantial reduction in retention time. The reduction in retention time (and hence run time) is usually worth the small sacrifice in efficiency.

Another reason for operating at OPGV is because the carrier gas linear velocity does not remain constant throughout the course of a temperature programmed run. As the column temperature rises, the viscosity of the carrier gas increases. As the carrier gas viscosity increases, there is a corresponding decrease in the carrier gas linear velocity (except when using Megabore direct injectors without a septum purge function). If the average

$$h_{min} = A + \frac{B}{\bar{u}} + C\bar{u}$$

h_{min} = height equivalent to a theoretical plate
$\bar{u}$ = average linear velocity
A   = multi-path flow term
B   = longitudinal diffusion term
C   = resistance to mass transfer term

Equation 1: Van Deemter Equation

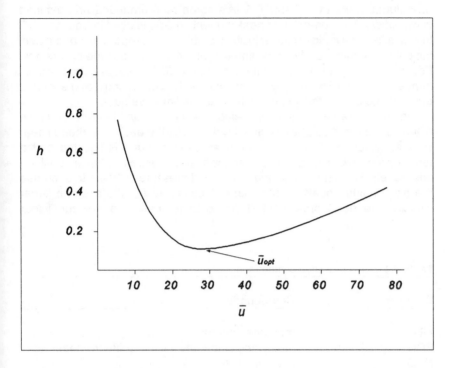

Figure 6-1: Typical Van Deemter Curve

linear velocity is set at $\bar{u}_{opt}$, the linear velocity decreases as the temperature program progresses to higher temperatures and the linear velocity proceeds to shift into a region of lower efficiency (left side of the Van Deemter curve). If the average linear velocity is set slightly above $\bar{u}_{opt}$, as the velocity decreases it shifts towards the minimum in the Van Deemter curve and into the region of higher efficiency.

6.4   Carrier Gas Measurements
6.4.1 Average Linear Velocity

The average linear velocity is determined from the retention time of a non-retained compound. Inject 1-2 µL of an appropriate compound using a split injector; dilution of the compound is necessary for a Megabore direct or on-column injectors.   Table 6-1 lists some recommended non-retained compounds. The non-retained peak's retention time is used in Equation 6-2 to calculate the average linear velocity. To obtain the proper units of cm/sec, the column length must be in centimeters and the retention time in seconds. For some very thick film columns and PLOT columns, use higher temperatures (50-75°C) to avoid retention of the compounds, and ensuring errors in the calculated values of the average linear velocity.

In practice, an average linear velocity is selected and the retention time of the non-retained peak is calculated which gives the desired average linear velocity (Equation 6-3).   The non-retained compound is injected and a retention time is obtained.   The carrier gas head pressure is adjusted until the desired retention time is obtained.   The head pressure is raised to decrease the non-retained peak retention time, thus increasing the average linear velocity; the head pressure is lowered to decrease the average linear

Table 6-1:   Non-retained compounds

Detector	Compounds
FID	methane, butane[1]
ECD	methylene chloride[2,3], dichlorodifluoromethane[2,3]
NPD	acetonitrile[3,4]
TCD, MS	methane, butane, argon, air

1.   A disposable lighter is a good source of butane. Place the syringe needle into the flame outlet of the lighter. Depress the small button allowing the gas to escape. Pull up 1-2 µL of the gas and inject.

2.   Headspace or diluted in solvent.

Headspace: Fill an autosampler vial with about 10 drops of solvent. Tightly seal the vial with a cap. Shake the vial for several seconds. Pierce the septum with the syringe needle, but do not insert the needle into the liquid. Pull up 1-2 µL of the headspace above the liquid and inject.

Solvent dilution: Add 25-50 µL of the appropriate solvent to 10 mL hexane or iso-octane and thoroughly mix. Inject 1-2 µL. Further dilution may be necessary depending on the sensitivity of the detector.

3.   Use a column temperature of 40°C or greater.

4.   Use a column temperature of 90°C or greater.

$$\bar{u} = \frac{L}{t_m}$$

L = column length (cm)
t_m = retention time of a non-retained compound (sec)

Equation 6-2:   Average Linear Velocity ($\bar{u}$)

velocity. A retention time ±0.05 minutes of the calculated value is sufficient in most cases. It is difficult to adjust the head pressure to obtain the exact value. Small errors and changes in the column length used in the calculation render the calculated retention time slightly inaccurate. Table 6-2 lists the retention times of the non-retained compound peak for a range of average linear velocities and column lengths.

The average linear velocity is dependent on column temperature. For this reason, the same temperature must be used each time the carrier gas velocity is set for a particular method. Using a lower column temperature results in a faster average linear velocity; a higher temperature results in a slower average linear velocity. Usually the initial temperature of the program is used because it is the most convenient point. In practice any temperature can be used; however, a consistent temperature for measurement is the most important consideration.

$$t_m = \frac{L}{60\,\bar{u}}$$

t_m = retention time of a non-retained compound (min)
L = column length (cm)
$\bar{u}$ = desired average linear velocity (cm/sec)

Equation 6-3:   Non-retained Peak Retention Time to Obtain A Specific Average Linear Velocity

Table 6-2:   Non-retained Peak Retention Times (min)

u	Column Length (m)					
	12	15	25	30	50	60
10	2.00	2.50	4.17	5.00	8.33	10.00
15	1.33	1.67	2.78	3.33	5.56	6.67
20	1.00	1.25	2.08	2.50	4.17	5.00
25	0.80	1.00	1.67	2.00	3.33	4.00
30	0.67	0.83	1.39	1.67	2.78	3.33
35	0.57	0.71	1.19	1.43	2.38	2.86
40	0.50	0.63	1.04	1.25	2.08	2.50
45	0.44	0.56	0.93	1.11	1.85	2.22
50	0.40	0.50	0.83	1.00	1.67	2.00
55	0.36	0.45	0.76	0.91	1.52	1.82
60	0.33	0.42	0.69	0.83	1.39	1.67
65	0.31	0.38	0.64	0.77	1.28	1.54
70	0.29	0.36	0.60	0.71	1.19	1.43
75	0.27	0.33	0.56	0.67	1.11	1.33
80	0.25	0.31	0.52	0.63	1.04	1.25

6.4.2 Flow Rate

There are several methods of measuring flow rate and each one gives a slightly different value. The values are different because the flow rates being measured are actually at different points in the column. The most common method of measuring flow rate is to use a flowmeter attached to the detector. This provides the flow rate at the exit or back of the column. Flow rates of 1-3 mL/min are used for small diameter ($\leq$0.32 mm I.D.) columns. These low flows are difficult to accurately measure with a flowmeter. For Megabore columns (0.53 mm I.D.), a low volume flowmeter usually gives accurate values since flow rates of 4-10 mL/min are common. Flowmeter inaccuracies at low flow rates and difficulties in obtaining a leak free connection with the detector create errors. Also, all detector and makeup gases need to be completely turned off to obtain a correct column flow rate. Note that bubble and digital flow meters give slightly different values due to the different mechanisms used to measure the flow rates.

Using a flow meter at the detector provides the column flow at the exit of the column. This flow is useful when the carrier gas flow and its impact on detector behavior are important (e.g., when using concentration dependent detectors). For accurate split ratio calculations and injector efficiency issues, the carrier gas flow rate at the front or entrance to the column is more important. This flow rate is lower than the flow at the exit of the column. It can be calculated using Table 6-3. All that is needed is the flow at the exit of the column, the ambient pressure and the column head pressure (from the gauge on the GC).

If a flow meter is unavailable, the average flow rate of carrier gas can be calculated using Equation 6-4. The retention time of a non-retained compound is the same as the one used in the average linear velocity

Table 6-3:   Carrier Gas Flows at the Front of the Column

Pressure Ratio [1]	Multiplier[2]
1.1	0.95
1.2	0.91
1.3	0.87
1.4	0.83
1.5	0.79
1.6	0.76
1.7	0.73
1.8	0.70
1.9	0.67
2.0	0.64
2.1	0.62
2.2	0.60
2.3	0.58
2.4	0.56
2.5	0.54

1. Ratio of the pressure at the front of the column (column head pressure + ambient pressure) and the pressure at the exit of the column (ambient pressure).

2. Multiply the measured carrier gas flow at the exit of the column by this number to obtain the flow at the front of the column.

calculation (Equation 6-2), but it is in minutes not seconds. The flow rate calculated using Equation 6-4 provides an average value and not the flow rate at any specific point in the column. It is not possible to directly measure the column flow into the detector for GC/MS systems; therefore, the average flow rate is often used.

Like linear velocity, flow rate is dependent on column temperature. For this reason, the same temperature must be used each time the carrier gas flow rate is measured or set for a particular method. Using a lower column temperature results in a faster flow rate; a higher temperature results in a slower flow rate. Usually the initial temperature of the program is used because it is the most convenient point. In practice any temperature can be used; however, a consistent temperature for measurement is the most important consideration.

$$\bar{F} = \frac{\pi\, r^2\, L}{t_m}$$

r = column radius (cm)
L = column length (cm)
t_m = retention time of a non-retained compound (min)

Equation 6-4:   Average Flow Rate

6.4.3   Converting Between Average Linear Velocities and Flow Rates

Occasionally, converting between average linear velocity and flow rate is desired. It is not easy to convert from average linear velocity to a column (exit or entrance) flow rate without a variety of equations and calculations. However, it is possible to easily convert between average linear velocity and average column flow using Equations 6-5a and b.

a)

$$\bar{u} = \frac{\bar{F}}{189\, r^2}$$

b)

$$\bar{F} = 189\, r^2\, \bar{u}$$

r = column radius (cm)
$\bar{u}$ = average linear velocity (cm/sec)
$\bar{F}$ = average flow rate

Equation 6-5:   Converting Between Average Linear Velocity ($\bar{u}$) and Average Flow Rate ($\bar{F}$)

6.5 Carrier Gas Selection

Hydrogen, helium, nitrogen and argon/methane are used as carrier gases in gas chromatography. Helium and hydrogen are the best ones for capillary columns. Nitrogen is a suitable carrier gas, but it is not recommended except for special cases. Argon/methane should not be used as a carrier gas for capillary columns.

6.5.1 Nitrogen

Nitrogen provides the best efficiency when compared to helium and hydrogen; however, $\bar{u}_{opt}$ is at a very low average linear velocity (Figure 6-2). Longer retention times occur at such low linear velocities. Substantial analysis speed is sacrificed for maximum efficiency. The Van Deemter curve for nitrogen is steep. This means efficiency drops sharply with any deviation away from $\bar{u}_{opt}$, thus the linear velocity range for maximum efficiency is small. Attempts to decrease run times by increasing the carrier gas average linear velocity causes a large loss of efficiency. Small errors in velocity measurements result in significant changes in efficiency.

The Van Deemter curve becomes steeper and the h_{min} value at u_{opt} increases as compound retention goes up. The large loss of efficiency as compound retention increases makes nitrogen a poor carrier gas for analyses where the compounds elute over a wide temperature range. The linear velocity decreases as the temperature program progresses. Many of the compounds are chromatographing while the carrier gas is well outside the velocity range of maximum efficiency.

Nitrogen is not recommended as a carrier gas for capillary columns primarily due to the low average linear velocity range and the resulting long run times. Also, the average linear velocity can only be optimized over a small temperature or retention range. For isothermal temperature conditions, short temperature programs and compounds that elute over a small range, nitrogen is a satisfactory carrier gas. For simple analyses (i.e., only a few compound eluting close together), the best efficiency is obtained with nitrogen. Nitrogen is the least expensive carrier gas. It is 30-50% less costly than helium and 10-25% less than hydrogen for equivalent purity grades.

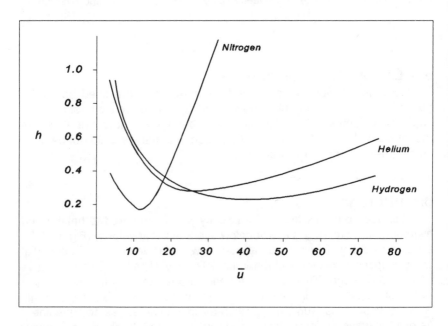

Figure 6-2:   Representative Van Deemter Curves for Nitrogen, Helium and Hydrogen

Figure 6-3: Comparison of Resolution and Run Time with Nitrogen, Helium and Hydrogen Note the differences in the retention times.
Column: DB-5, 15 m x 0.25 mm I.D., 0.25 µm
Oven: 55°C for 1 min, 55-155°C at 25°/min
155-300°C at 10°/min, 300°C for 20 min
Compounds: PAH's

6.5.2 Helium

The Van Deemter curve for helium shows that $\bar{u}_{opt}$ is at a higher average linear velocity than for nitrogen (Figure 6-2). The h_{min} value at $\bar{u}_{opt}$ is slightly higher than nitrogen, but the difference is small. The higher $\bar{u}_{opt}$ value allows the use of higher linear velocities without sacrificing a large amount of efficiency. In general, helium provides nearly equivalent resolution to nitrogen, but at much shorter retention times (Figure 6-3).

The Van Deemter curve for helium is not as steep as for nitrogen. The flatter curve results in small changes in efficiency as the column temperature changes in a temperature program. The carrier gas linear velocity remains in the range of higher efficiency throughout the temperature program run. There is only a small loss of efficiency as compound retention increases. This makes helium a good carrier gas for compounds that elute over a wide temperature or retention range.

Helium is the most expensive carrier gas. It is difficult to obtain at a reasonable price in many parts of the world. If helium is not readily available, hydrogen should be used.

6.5.3 Hydrogen

Hydrogen is the best carrier gas for capillary columns. Its Van Deemter curve is very flat and $\bar{u}_{opt}$ is at very high average linear velocity (Figure 6-2). The h_{min} value at $\bar{u}_{opt}$ is slightly higher than nitrogen and equivalent to helium. The very high $\bar{u}_{opt}$ value allows the use of very high linear velocities without sacrificing a large amount of efficiency. Shorter retention times are obtained compared to helium and especially nitrogen while maintaining high efficiency (Figure 6-3). The extreme flatness of the hydrogen Van Deemter curve makes hydrogen vastly superior for compounds eluting over a wide temperature or retention range.

Hydrogen is less costly than helium (10-25% less) but more costly than nitrogen. There are several hydrogen generators commercially available that produce high quality hydrogen suitable for use as a carrier gas. This eliminates the purchasing and changing of high pressure cylinders for high volume users of hydrogen. Most hydrogen generators return their original purchase cost in 2-3 years.

There is a reluctance to use hydrogen as a carrier gas due to its flammability. This a valid concern, but the actual hazards are not as severe as commonly perceived. A flammable gas and the high temperatures of the GC may seem like a dangerous combination. However, there are a variety of reasons that make working with hydrogen less hazardous than it seems.

The amount of hydrogen in air must be around 4% before the possibility of an explosion or sustained combustion exists. An explosion will not occur if the value is slightly higher or lower. Hydrogen is very diffusive in air, making it very difficult to obtain a high enough buildup of hydrogen to make an explosive mixture with air. Ventilated rooms and the GC fan clear the hydrogen away before it can build up. A completely closed environment and delivery of a very high volume of hydrogen are necessary to build up enough hydrogen for a problem to occur. It usually requires the GC oven fan to be off and a specific type of injector design.

Some GC's are equipped with pressure regulators (i.e., forward pressure regulated injectors) to control the total amount of carrier gas entering the GC. This type of system adjusts the flow of carrier gas to maintain the pressure as set on the head pressure gauge. If the flow path is exposed to the ambient pressure (e.g., a leak or broken column), the pressure controller attempts to pressurize the GC oven to the set pressure. A substantial volume of hydrogen is delivered in a short amount of time. Nearly every incident of a hydrogen explosion has been with a forward pressure regulated injector system and some type of gross operational error. Other GC's are equipped with flow controllers (i.e., back pressure regulated injectors) to control the total amount of carrier gas entering the GC. In the worst case, only 100-200 mL/min of hydrogen can be delivered to the GC. Even with a leak directly into the GC oven, too little hydrogen is supplied to reach an explosive level. As an additional safety feature, all modern GC's are designed with an explosion absorbing oven door. If an explosion does occur, the door absorbs most of the force of the explosion and it opens instead of becoming a dangerous projectile.

If hydrogen explosions are still a concern, simple installation of flow controllers in the gas supply lines reduces the possibility of a hydrogen explosion due to leaks into the oven (Figure 6-4). Each flow controller is set 10-20% above the total volume of hydrogen needed from that hydrogen line. If there is a leak downstream of the flow controller, only the gas volume set with the controller is delivered. This will prevent a buildup of hydrogen at the leak site.

6.6   Recommended Linear Velocities

Unless the compounds all elute within several minutes of each other, there is rarely a single average linear velocity that is the best for all of the compounds in a sample. The optimal linear velocity is also influenced by column dimensions. Table 6-3 lists the average linear velocity range

Figure 6-4: Hydrogen Line Flow Controllers

Table 6-3:   Recommended Average Linear Velocity and OPGV

Nitrogen

Diameter (mm)	Length (m)	Film (µm)	$\overline{u}$	OPGV
All	All	All	10-20	15-25

Table 6-3: Recommended Average Linear Velocity and OPGV (Cont.)

Helium

Diameter (mm)	Length (m)	Film (µm)	$\bar{u}$	OPGV
0.20	12	0.33	45-55	60-85
0.20	25	0.33	30-40	35-60
0.20	50	0.33	25-35	30-45
0.25	15	0.25	40-50	50-80
0.25	30	0.25	30-40	40-70
0.25	60	0.25	25-35	30-50
0.25	15	1.0	35-45	50-80
0.25	30	1.0	30-40	35-65
0.25	60	1.0	25-35	30-50
0.25	15	3.0	25-35	40-75
0.25	30	3.0	20-30	35-60
0.25	60	3.0	15-25	30-50
0.32	15	0.25	40-50	50-80
0.32	30	0.25	30-40	40-70
0.32	60	0.25	25-35	30-50
0.32	15	1.0	35-45	50-80
0.32	30	1.0	30-40	35-65
0.32	60	1.0	25-35	30-50
0.32	15	3.0	25-35	40-75
0.32	30	3.0	20-30	35-60
0.32	60	3.0	15-25	30-50
0.53	15	1.0-1.5	30-40	50-85
0.53	30	1.0-1.5	25-35	35-70
0.53	60	1.0-1.5	20-30	25-55
0.53	15	3.0	20-30	50-85
0.53	30	3.0	20-30	40-75
0.53	60	3.0	15-25	20-50

Table 6-3: Recommended Average Linear Velocity and OPGV (Cont.)

Hydrogen

Diameter (mm)	Length (m)	Film (µm)	$\overline{u}$	OPGV
0.20	12	0.33	70-80	85-150
0.20	25	0.33	50-60	70-105
0.20	50	0.33	40-50	50-75
0.25	15	0.25	60-70	75-130
0.25	30	0.25	50-60	70-110
0.25	60	0.25	40-50	60-80
0.25	15	1.0	50-60	85-130
0.25	30	1.0	45-55	65-110
0.25	60	1.0	40-50	45-80
0.25	15	3.0	30-40	75-150
0.25	30	3.0	25-35	60-85
0.25	60	3.0	20-30	50-65
0.32	15	0.25	60-70	75-130
0.32	30	0.25	50-60	70-110
0.32	60	0.25	40-50	60-80
0.32	15	1.0	50-60	85-130
0.32	30	1.0	45-55	65-110
0.32	60	1.0	40-50	45-80
0.32	15	3.0	30-40	75-150
0.32	30	3.0	25-35	60-85
0.32	60	3.0	20-30	50-65
0.53	15	1.0-1.5	40-50	80-150
0.53	30	1.0-1.5	40-50	70-115
0.53	60	1.0-1.5	35-45	60-100
0.53	15	3.0	25-35	75-125
0.53	30	3.0	25-35	65-100
0.53	60	3.0	20-30	55-100

recommended to obtain maximum efficiency and the OPGV range for the most common column sizes. The nitrogen Van Deemter curve is so sharp that the linear velocity differences between different column dimensions are small; therefore, the same linear velocity range applies to all columns.

If maximum efficiency is desired, some experimentation using different linear velocities within the recommended ranges is required. In most cases, very small efficiency differences are obtained for the average linear velocities within the range. Using a consistent average linear velocity for a given analysis is more important than worrying about the exact value.

6.7 Gas Purities

The recommended grades of carrier gas are normally called ultra high purity and oxygen free. These grades typically contain less than 1-2 ppm oxygen and 1 ppm hydrocarbons. The most important gas quality criterion is a very low concentration of oxygen. Oxygen rapidly degrades stationary phases and thus reduces column life. Low hydrocarbon contaminants is desirable and beneficial for most detectors. Several ppm of inert gases rarely have any noticeable effect. Chromatographic, research and similar grade gases are not necessary. They are extremely pure and their very high cost is without any real benefits.

Gas impurity traps are often recommended to further purify carrier gases. The lower the grade of the carrier gas, the more important gas impurity traps become. The use of traps has to be balanced against the cost of the traps and the higher cost of better grades of carrier gas. Lower grade, thus lower cost carrier gases cause the traps to expire more rapidly. The savings in carrier gas cost is usually offset by the higher cost of frequent trap replacement. Better grades of gas do not cause the traps to expire as quickly. The higher cost of the better grade gas is nearly offset by the lower cost of less frequent trap replacement.

The quality of many ultra high purity and oxygen free carrier gases is quite high. Recently, it has been noted that the use of gas impurity traps do not result in substantial improvement in the cleanliness of the carrier gas. Unless performing trace level analyses or using a detector at its maximum sensitivity level, the use of gas impurity traps with high quality carrier gases is rarely worth the cost and required periodic replacement efforts.

6.8   Common Carrier Gas Problems

There are only a few problems that are associated with the carrier gas. The most difficult one to isolate and solve is a gas contamination problem.

Contaminated gas (i.e., a bad cylinder of gas) is fairly rare. The carrier gas passes through many areas. Volatile materials in any of these areas can be transported by the carrier gas into the column where they are detected as a problem in the chromatogram. The only easy method to determine whether the carrier gas is contaminated is to change the cylinder. If the background problem does not substantially decrease or disappear in several hours, the GC is contaminated and the carrier gas is not the problem. If the problem disappears, the carrier gas may be at fault. It may be a good idea to re-install the original cylinder if that carrier gas seems to be the source of the problem. Sometimes problems disappear just because something was changed. If the problem returns, the gas cylinder is at fault; if the problem does not return, something else was responsible for the problem. Gas cylinders should not be used below 100-200 psi (700-1400 kPa). As the cylinder pressure decreases to 100-200 psi, contaminants in the cylinder begin to vaporize and contaminate the gas. At higher pressures, the contaminants are liquids and do not have enough vapor pressure to leave the cylinder.

Improper measurement or setting of the linear velocity results in efficiency changes and retention time shifts. If an efficiency changes occurs, but the retention times do not shift, the problem is not with the linear velocity. If the retention times shift all in the same direction, it is possible the incorrect linear velocity or temperature is being used. All compounds are equally affected by a change in the linear velocity. If the carrier gas velocity changes, the retention times change by the same amount (absolute not percentage). If a previously run chromatogram is available, calculate the partition ratios (k) for each peak and compare them to the current chromatogram. If the carrier gas velocity changed, the k values for both chromatograms will be the same. If the k values are not the same, stationary phase behavior or column temperature has changed. If the carrier gas head pressure does not exceed a particular pressure even though the column head pressure knob is turned up, there may not be enough gas supplied to the GC. Check the pressure of the incoming carrier gas line to make sure it is at the pressure recommended by the GC manufacturer. If the pressure is too low, not enough gas is supplied to the GC to deliver the desired flow. If the head pressure increases when the split line flow is decreased, this is the problem. If the incoming pressure is high enough, check for leaks in the carrier line especially within the GC. Finally, check the flow controller if the GC has a back pressure regulated injector. Many flow controllers have a flow element which limits the range of flows the flow controller can deliver. The flow controller can only deliver the maximum flow as governed by the flow element regardless of the incoming carrier gas pressure. The flow element needs to be changed to a higher flow version to obtain the higher flow rates.

Some flow controllers have an adjustable flow element instead of a replaceable one.

If the linear velocity changes by a large amount with a small turn of the head pressure controller knob, a flow element with a smaller range may be needed. One rotation of the controller knob changes the linear velocity by a smaller amount with a lower range flow element. Finer adjustments to the linear velocity are possible when the range of the flow controller is smaller. This is especially helpful when operating at low carrier gas head pressures.

7   INJECTORS

7.1   Introduction

The primary goal of sample injection is to introduce the sample into the column; however, there is much more to this operation. It is important to deposit the sample in the column in the narrowest band. The shorter the band at the beginning of the chromatographic process, the shorter it will be as it exits the column. This results in the desirable effect of tall and narrow peaks in the chromatogram.

One of the common assumptions about capillary GC is that the composition of the sample introduced into the column is identical to that of the sample before injection. Except for on-column injection, this assumption is incorrect.   The operating conditions, hardware choices and sample composition all influence the similarity between the injected and original sample. Fortunately, the proper use of standards renders the differences unimportant.   Several injection techniques have been developed to meet the specific demands of capillary gas chromatography. A universal injector or sample introduction approach has not been developed that can accommodate the wide diversity of capillary columns, samples and automation requirements. Accurate and precise injection results can be obtained if the various benefits and limitations of each injection method are understood and appreciated.

There are two basic types of injectors for capillary columns - vaporization and on-column. Vaporization injectors are more common, and thus more familiar to most analysts. Vaporization injectors include split, splitless and Megabore direct.   On-column injectors include the cool on-column and Megabore on-column. Regardless of the type of injector, each has specific advantages and operation requirements.

7.2   The Basics of Vaporization Injectors

All vaporization injectors for capillary columns function basically in the same manner. Each GC manufacturer has its own injector design with the two major designs being back pressure and forward pressure regulated. The difference between these two lies in the method of regulating the various gas flows. Appendix F describes the design and performance differences in detail.

All vaporization injectors have similar design features (Figure 7-1). A glass liner resides inside the heated, metal injector body. A carrier gas line supplies carrier gas to the interior of the injector body. Usually the carrier gas enters near the top of the injector. A syringe is used to pierce the septum and introduce the sample into the injector. The high temperature of the injector causes the volatile components of the sample to rapidly vaporize.

Figure 7-1:   Basic Vaporization Injector

In some cases, a gaseous sample may be directly introduced into the injector. Any vaporized sample mixes with the carrier gas and is carried anywhere the carrier gas travels. Sample is introduced into the column by the movement of the carrier gas into the column. After the sample enters the column, the separation process begins. If there are flow paths other than the column (i.e., gas lines) for the carrier gas, portions of the sample do not enter the column. The presence of these gas lines and the volume of carrier flowing through them is determined by the design and use of the different injectors. The flow of gas through each of the gas lines and column is controlled by flow controllers or pressure regulators.

Most capillary injectors have a septum purge feature. The septum purge is a low flow of carrier gas across the top of the injector immediately below the septum. This gas then leaves the injector via the septum purge line. Flows of 0.5-5 mL/min are usually recommended by GC manufacturers. Septum purge flows of 1-3 mL/min are the best overall settings. The septum purge helps to minimizes the condensation of non-volatile or high boiling materials on the exposed portion of the septum. This region of the injector can be more than two times cooler than the set injector temperature. Condensation of sample residues in these areas can result in injector contamination problems. The septum purge also helps to minimize septum bleed problems (Section 7.10.2). The septum purge helps to keep the injector clean and, if used properly, does not have a significant influence on sample vaporization or column introduction processes.

7.2.1 Injector Temperature

The injector temperature should be just hot enough to ensure "instant" vaporization of the entire sample. Excessively high temperatures may cause degradation of thermally unstable compounds. Also, massive expansion of the sample solvent and the more volatile sample compounds is always a concern at higher injector temperatures. This effect is called backflash (Section 7.2.3) and can result in a large solvent front, peak tailing or carryover problems.

Excessively low injector temperatures may cause incomplete or slow vaporization of the sample. This may result in broad or tailing peaks especially for compounds with high boiling points. The problem is less noticeable for injectors that transfer the sample to the column very rapidly. Injector contamination is a greater problem at lower injector temperatures. Larger amounts of the high boiling point sample components accumulate in the injector since they are not readily vaporized and carried into the column.

7.2.2 Speed of Sample Transfer

Rapid transfer of the vaporized sample from the injector into the column is desired. This results in the highest efficiency. Rapid sample transfer is obtained when the carrier gas sweeps the volume of the injector liner very quickly. High carrier gas flows and small liner volumes both result in rapid sweep rates. Rapid sample transfer creates an initial sample band that occupies a very short length of column, thus often resulting in peaks that are narrow. If the sample band is spread over a long length of column, the peaks may be excessively wide. The use of excessively high carrier gas flow rates or small liners to obtain high efficiencies can result in other problems. Other factors such as optimal carrier gas linear velocities and the reduction of backflash have to be considered.

Due to their design, some injectors have a low flow of carrier gas in the injector at the time of sample introduction. This results in a slow transfer of the sample into the column. In some cases, the loss of efficiency can be tolerated. In other cases, the wide sample band has to be focussed into a shorter band by the use of specific conditions or techniques. Some of the operational requirements for some injectors are necessary to overcome the poor (i.e., slow) sample transfer characteristics of that injector.

7.2.3 Backflash

Upon vaporization in the injector, a sample expands to many times its original volume. The force of this expansion can be quite large and often generates a momentary high pressure pulse greater than the carrier gas pressure in the injector. If the final volume of the vaporized sample exceeds the injector liner volume, some of the vapors flow out of the liner. This event is called backflash.

The more volatile a compound, the more likely it is to be susceptible to backflash. In most samples, the solvent is present in substantially greater amounts than the other sample components. Also, it is usually the most volatile sample component. Both of these factors makes the solvent the most likely substance to backflash. As the solvent flashes out of the injector liner, it may carry some of the other volatile sample components along with it.

As the vapors leave the liner, they may travel to a number of places. The vapors that leave the top of the liner flow across the bottom face of the septum and are either swept out through the septum purge line or back into the injector liner. If the pressure pulse is greater than the injector pressure, the vapors may travel into the incoming carrier gas line. The vapors that leave the bottom of the liner are swept out of the split vent line (if present or open).

The various compounds condense on any surface that is below their respective boiling points. These areas include the small portion of the septum exposed to the interior of the injector and the gas lines. Most gas lines are not heated and are at ambient temperature along most of their length. The highest boiling compounds deposit closest to the injector with the most volatile portions traveling the greatest distance. The most volatile compounds may travel over 10 cm before condensing inside the gas lines.

Any compounds condensed on the septum eventually re-enter the injector liner. This may occur in two ways. Small amounts of the compounds may slowly volatilize and be carried back into the injector. Depending on the column temperature, these compounds either contribute to the background noise or are evident as peaks. If the column is maintained at higher temperatures, increased background noise is most common; interfering or ghost peaks are evident if the column is maintained at lower temperatures. Unless the contamination is very large, the background noise does not have any significant negative impact. The other route for entering the injector is from the continuation of backflash. Solvent vapors from any backflashing solvent passes over the septum and dislodges some of the condensed compounds. If the vapors are carried back into the injector, these compounds then chromatograph just like an injected sample, resulting in interfering or ghost peaks. The larger the amount of condensed compounds, the greater the problem caused by the background noise or extra peaks.

Any sample that has condensed in the septum purge or split vent lines normally does not cause any significant problems unless the contamination is very severe. Contamination of the incoming carrier gas line poses the same problem as compound condensation on the septum. Small amounts of the contaminants can be carried into the injector by the carrier gas. Also, sample solvent entering the gas line from backflash can carry contaminants back into the injector, resulting in background noise or extra peaks.

In addition to injector contamination problems, backflash can also cause peak tailing or broadening. Since some of the sample leaves the liner, additional time is now required for the carrier gas flow to sweep that portion of the sample back into the injector liner and column. The sample is introduced into the column in a much broader band (i.e., longer period of time). If only a small portion of the sample left the liner, peak tailing is usually evident. If a large portion of the sample left the liner, peak broadening is more prevalent. The peak tailing or broadening usually becomes less severe as peak retention increases. This is due to the greater backflashing of the more volatile sample compounds and focussing at the front of the column of the less volatile compounds.

The area outside of the injector liner is primarily stainless steel. Compounds with hydroxyl or amine groups are particularly sensitive to interaction with heated metal surfaces. Peak tailing or loss of peak size may be evident if a compound interacts with a metal surface. For these cases, it may be especially important to eliminate or minimize any backflash.

The amount of sample expansion primarily depends on the sample solvent, injector temperature and column head pressure. Table 7-1 lists some common solvents and the approximate amount of sample expansion. If backflash is a problem, the use of a sample solvent with a low expansion volume helps minimize the problem.

Table 7-1:   Solvent Expansion Volumes

Solvent	Approximate Volume (µL) of Vapor	
	1 µL	2 µL
i-Octane	110	220
n-Hexane	140	280
Toluene	170	340
Ethyl acetate	185	370
Acetone	245	490
Methylene chloride	285	570
Carbon disulfide	300	600
Acetonitrile	350	700
Methanol	450	900
Water	1010	2020

Injector Temperature: 250°C   *Pressure:* 20 psig

7.2.4 Discrimination

It is common with vaporization injectors for the entire sample not to be fully vaporized at the time of sample introduction into the column. Vaporization becomes faster and more complete as a compound's volatility increases. Since only the vaporized portion of the sample enters the column in significant amounts, a different percentage of each compound enters the column. A greater percentage of the more volatile compounds enters the column than the less volatile compounds. This effect is called injector discrimination. It is usually evident as progressively decreasing peak size for the later eluting compounds in the sample (Figure 7-2).

Figure 7-2: Example of Injector Discrimination
Column: DB-5, 15 m x 0.25 mm I.D., 0.25 µm
Injector: Split 1:100
Oven: 90°C for 1 min, 90-225°C at 15°/min
Carrier: Helium at 32 cm/sec
Sample: n-C_{10}-C_{16} evens in hexane at 500 ng/µL each

There are a number of factors that influence the amount of injector discrimination. The main factors being the sweep time of the carrier gas, amount of mixing of the vaporized sample prior to column introduction and the volatility range of the compounds. The faster the sample is swept into the column, the less time that is allowed for vaporization. The more volatile compounds have enough time to become substantially vaporized while the less volatile compounds do not; therefore, greater amounts of the more volatile compounds enter the column. This makes discrimination greater at high carrier gas sweep rates. Liners designed to mix the vaporized sample prior to introduction into the column aid in reducing the amount of discrimination. Greater amounts of sample mixing result in lower discrimination. These types of liners are primarily useful for high sweep rate situations.   The larger the volatility differences between the sample compounds, the greater the extent of discrimination. Later eluting peaks can be over 10 times smaller than earlier eluting peaks in a chromatogram when the compounds have boiling points differing by 200-300°C.

Factors including injector temperature, injection size and technique, and injector design have minor influences on the amount of discrimination. The magnitude of their influence is unique to each sample and can be difficult to predict. Usually the amount of discrimination influenced by these factors is small and rarely considered when selecting operating conditions.

7.3   Split Injectors
7.3.1 Description

The concentrations of many samples easily exceed the capacities of most conventional capillary columns.   Limiting the amount of sample reaching the column by splitting it prior to column introduction prevents column overloading; a split injector performs this function. The injector splits the sample into two unequal portions with the smaller fraction going to the column and the larger fraction being discarded. Since only a very small portion of the sample enters the column, thus reaching the detector, split injectors are used for concentrated samples. Trace level analyses are rarely performed using a split injector. Typical sample concentrations for split injections are usually 0.1-10 µg/µL per sample component.

A split injector consists of the heated injector, liner, incoming carrier gas line, split line and septum purge line (Figure 7-3). After entering the injector, the carrier gas exits by way of the column, split line and septum purge line. Since the septum purge flow is very low and above the region where sample vaporization occurs, very little sample is lost via the septum purge line. Any sample loss from the septum purge line is usually ignored when considering sample splitting.

After injection and vaporization, the sample vapors are mixed with the carrier gas. The vaporized sample travels wherever the carrier gas goes. The carrier gas can travel down two pathways - the column and the split line. The volume of carrier gas entering the column is small compared to the split line, thus only a small fraction of the sample enters the column. This splitting of the carrier gas flows results in the splitting of the sample also.   The amount of splitting is measured by the split ratio.

Figure 7-3:   Split Injector
Some details are omitted for clarity.

7.3.2 Split Ratio

The split ratio is the volume of carrier gas entering the column versus the volume leaving via the split line. The ratio of these two gas volumes is called the split ratio. The split ratio is determined by measuring or calculating the gas flows out the split vent and column. Split ratios are reported as the column flow normalized to one. There are two slightly different viewpoints on the measurement of the split ratio, and thus two equations (Equations 7-1a and b). The differences are small and of no consequence.

a: Ratio of amount of sample in the column and total sample

$$Split\ Ratio = \frac{F_{column} + F_{split\ vent}}{F_{column}}$$

b: Ratio of amount of sample in the column and the amount leaving via the split line

$$Split\ Ratio = \frac{F_{split\ vent}}{F_{column}}$$

$F_{split\ vent}$ = Flow from split vent (mL/min)
F_{column} = Column flow (mL/min)

Equation 7-1:   Split Ratio Calculations

The volume of carrier gas sweeping sample towards the column is the sum of the column and split line flows. This high volume of carrier gas results in a very rapid sample transfer into the column, thus high injector efficiencies. Most split injectors can deliver up to approximately 300 mL/min of total gas flow. The highest possible split ratio is limited to situations were the sum of the column and split flows do not exceed about 300 mL/min. The actual flow values depend on the model and setup of the GC. Small diameter columns require very low carrier gas flows, thus very high split ratios are obtainable. Large diameter columns use much higher gas flows, and the highest possible split ratios are somewhat lower. The lowest total flow in the injector (column + split line) should be about 20 mL/min. Lower values result in slow injector sweep rates, thus poor injector efficiencies (i.e., broad peaks). Large diameter columns use high carrier gas flows, thus low split line flows (resulting in low split ratios) can be used without dropping the total flows below 20 mL/min. Small diameter columns use low carrier gas flows, and the split line flows therefore need to be high to keep the total flow above 20 mL/min. The lowest possible split ratios for small diameter columns are higher than for a larger diameter column. Recommended split ratios for various column diameters are shown in Table 7-1.

Table 7-1:   Lowest Recommended Split Ratios

Column Diameter (mm)	Lowest Split Ratio
0.18-0.20	1:20 - 1:25
0.25	1:15 - 1:20
0.32	1:10 - 1:12
0.53	1:3 - 1:5

The split ratio selected is primarily determined by the amount of sample that needs to be introduced in the column. Low split ratios introduce a greater amount of sample into the column than a higher split ratio. Lower split ratios are used when sample concentrations are low and/or detector sensitivity is lacking. Higher split ratios are used for highly concentrated samples that may overload the column or detector. There are two other factors that need to be considered when selecting a split ratio - injector efficiency and discrimination.

At high split ratios, the flow of carrier gas in the injector is very high. This sweeps the sample from the injector and into the column within a very short time. Since the sample is introduced into the column so quickly, high injector efficiency is obtained. Also, at high split ratios, the high boiling (less volatile) sample compounds have very little opportunity to completely vaporize during their residence time in the injector. The low boiling (more volatile) compounds are more thoroughly vaporized within the same time period. The difference in the amounts of high and low boiling compounds introduced into the column can be large. This results in high injector discrimination. At low split ratios, the flow of carrier gas in the injector is lower than for high split ratio situations. The residence time of the sample in the injector is much longer than at a higher split ratio. The higher boiling compounds have a better opportunity to vaporize more thoroughly.   The lower boiling compounds vaporize so rapidly that longer times in the injector do not substantially increase the amount vaporized. The difference in the amounts of low and high boiling compounds introduced into the column is much smaller. This results in lower injector discrimination. In general, injector discrimination increases as the split ratio becomes larger.

A proper balance between amount of sample introduced into the column, discrimination and efficiency must be reached. Using a low split ratio places more sample in the column and minimizes discrimination; however, efficiency losses may be evident. Using a high split ratio places less sample in the column and discrimination is more severe; however, greater efficiency is gained. In most cases, the amount of sample introduced into the column is the most important factor. Losses or gains in efficiency with changes of split ratio are usually small and are not noticed unless large split ratio changes are made. Discrimination differences are noticeable when the split ratio is changed, but the effect is counteracted by the proper use of calibration standards.

A common assumption is that the sample is split in the same proportions as the split ratio indicates. Due to discrimination and errors in flow measurements, the actual amount of sample introduced into the column is not exactly what the split ratio indicates. At a 1:50 split ratio, it seems that 1 part of the sample enters the column for every 50 parts discarded via the split line. The actual amount of each compound entering the column deviates from the expected 1 to 50 ratio depending on a large number of variables. These variable include compound volatility, injector temperature and design, type of liner, column diameter, carrier gas flow, column position in the injector, injection size, and sample solvent. If all of these variables remain constant, very reproducible splits will be obtained. Each compound may be affected differently by change to any of these variables.

7.3.3 Septum Purge for Split Injectors

The septum purge line can act as a miniature split line if sample reaches the top of the injector during the vaporization and expansion process. The more volatile sample components more readily vaporize and expand into a greater volume. This makes them more likely to travel up into the septum purge gas stream and be carried out of the injector. Fortunately, the rapid sweeping of the sample into the column, even for low split ratio situations, does not permit much time for the sample to remain at the top of the injector. Since the sample spends little time in the region of the septum purge flow, only a small fraction of the sample is lost. For this reason, higher septum purge flows (e.g., 3-4 mL/min) can be used with split injector with minimal loss of the more volatile sample components. In any case, a greater loss of sample occurs for large injection volumes, high injector temperatures, small volume liners, and low carrier gas flows.

7.3.4 Split Injector Liners

There are many liners available for split injectors. The primary function of any liner is to provide an inert environment for sample vaporization. It is important to use a liner that is deactivated or active compounds may absorb onto the liner surface. Peak tailing or loss of peak size may be evident if absorption is occurring. Most liners are supplied already deactivated, thus new liners are suitable for immediate use. Liner deactivation decreases with use especially after injections with strongly acidic or alkaline samples. Liner activity is much greater problem for compounds containing hydroxyl or amine functional groups. Compounds lacking these groups are rarely affected. Liner activity is not a major problems for split injectors since the vaporized sample is swept into the column (and out of the injector) so quickly. There is little opportunity for adsorption to occur in such a short time period.

Split liners are available in a variety of designs ranging from simple, straight tubes to ones with complex inner surfaces; some may be packed with glass wool. Split liners with complex inner surfaces are designed to thoroughly mix the vaporized sample before it is split. The carrier gas mixed with vaporized sample does not flow directly to the bottom of the liner but is mixed in a turbulent fashion. This extensive amount of mixing helps to minimize discrimination by making the sample more homogeneous. Liners that reduce discrimination can be especially useful for high discrimination situations such as high split ratios or samples with a broad range of compound volatilities. Greater discrimination is seen with liners without some type of mixing feature. The inverted cup, "cyclo", and fritted liners are examples of liners that exhibit lower discrimination.

Some liners are packed with glass wool. The glass wool must be silylated or severe activity problems (i.e., peak tailing or size reduction) occur. Since it is very difficult to silylate glass wool, it is strongly recommended to purchase pre-silylated glass wool. A small amount of glass wool placed in the proper position reduces discrimination especially for the straight tube liners. The carrier gas has to travel through the bundle of glass wool which causes a thorough mixing of the vaporized sample and carrier gas. This mixing, just like for the complex surface liners, reduces the amount of discrimination. The glass wool also acts as a filter to minimize the amount of non-volatile materials reaching the column, thus reducing the possibility of contamination. The use of glass wool comes with several precautions. The plug of glass wool should not be too tightly packed. The carrier gas has difficulty flowing through the dense plug of glass wool. This causes the sample to reside in the injector too long with a subsequent broadening of the peaks. The pressure of the carrier gas may push a dense glass wool plug towards the bottom of the liner. A plug of glass wool at the bottom of a liner usually causes irreproducible peak size, blockage of the column or split line, or distorted peak shapes (i.e., tailing, rounded, broad or split). When forming the plug of glass wool, breakage of the individual fibers should be minimized. While the surface of the glass wool is deactivated, the interior is not. Any ends created upon fiber breakage are potential active sites. The glass wool plug should be formed by gently rolling the fibers into a loosely packed ball about the size of the liner's inner diameter. Excessive mashing or crunching of the glass wool plug into a ball creates numerous ends. Conversely, a glass wool plug that is excessively loose does not mix the carrier gas enough to reduce discrimination or trap non-volatile materials. The plug of glass wool should be gently and carefully pushed into the proper position in the liner. For maximum injection reproducibility, the same person should install the glass wool or liners with pre-installed glass wool should be purchased. Surprisingly, even small differences in the amount of glass wool and how tightly it is packed affect the degree of discrimination.

There are differing opinions and studies concerning the best position for glass wool in a split liner. The most common conclusion is to have the glass wool below the end of the syringe needle. Other studies have shown better results if the needle passes through the glass wool. It is suspected that the glass wool wipes residual sample from the needle and provides better injection reproducibility. In any case, the most important factor is consistent placement of the glass wool.

Probably, the most important aspect of selecting a split injector liner is consistency. Each type of liner provides different results especially when discrimination is considered. If maximum reproducibility from analysis-to-analysis or lab-to-lab is desired, the same type of liner needs to be used each time. There are many claims stating that a specific liner is especially good for a certain type of sample. While this may be true, there are so many variables that affect discrimination, linearity, sensitivity, efficiency and peak shape. Liners are often only one of many variables that influence the final results. In general, liners with the complex inner surfaces are more costly. They will often cost 2-4 times more than a straight tube or very simple liner. This may be an important consideration if the liners need frequent replacement. If liners are re-used, the complex liners are more difficult to clean and properly re-silylate. In either case, the more costly or difficult to clean liners may not be worth the reduced discrimination they provide. Essentially, a good liner is one that provides satisfactory results.

7.3.5 Column Position in Split Injectors

Each model of GC has a distance to which the column is inserted into the injector. This value is usually found in the column installation instructions for the GC. The distance recommended by the GC manufacturer is probably the best one for the widest variety of samples; however, it may not be the best one for a specific sample. Errors in the insertion distance, especially if the distance is too short, may result in severe problems. The column insertion distance primarily influences efficiency and discrimination. Sometimes, even several millimeters difference has a pronounced impact on these parameters. The best distance varies depending on the nature of the sample and the specific liner.

The GC manufacturer's usually have more than one liner available for each GC. Also, there may be a number of different liners available from other sources. The manufacturer's column insertion distance may not be suitable or optimal for some of these other liners. This is due to the position of any flow disruption devices or an unique design. In general, the end of the column should be near the bottom of the liner. It is quite important that the column end is not completely below the liner and in the ferrule or fitting area of the injector. If the column end resides in this area, poor peak shape, reduced peak size or severe discrimination may occur.

Regardless of the type of liner, the position of the column in the liner may have a significant effect on efficiency and discrimination. Experimenting with column position, within the previously discussed guidelines, may result in the best efficiency or minimal discrimination for a particular sample. If optimizing these two parameters is not extremely important, using the GC manufacturer's recommended insertion distance is satisfactory. Most importantly, consistent column position is critical if precise reproducibility is required.

7.3.6 Other Aspects of Split Injectors

Split injectors use high volumes of carrier gas especially at high split ratios. It may be desirable to use a slightly lower split ratio when possible to save on gas consumption. For example, dropping to a 1:50 split ratio from 1:100 reduces gas consumption by half. Approximately two times the sample enters the column (which in many cases is not a problem), but a substantial savings in carrier gas costs results. Devices to reduce split line flows and return them to the original flows with a simple turning of a switch are available. These are useful to reduce carrier flows during periods of non-use; and they may quickly return the investment by the savings on gases.

Small traps are available for the split vent line. These are designed to trap any of the compounds that leave the split vent. Sample solvent comprises most of the material leaving the split vent. Most sample components condense somewhere along the split line which is usually at room temperature. These split line traps primarily help to keep volatile compounds such as solvents from contaminating the lab air. Depending on their structures, some of these compounds can be detected by sensitive GC systems as part of the background. For example, methylene chloride contamination of the lab air is a common problem for some laboratories performing low level halocarbon analyses.

Split line traps can become clogged or plugged with use. This creates back pressure in the split line. This may change split ratios and alter column carrier gas flows in some systems. Inspection and replacement of these traps is important. The split line can become heavily contaminated from the condensation of sample in this cool area. Occasional cleaning of the split line is recommended to prevent these materials from eventually coming out the split vent or backwashing into the injector.

7.3.7 Common Problems with Split Injectors

Split injectors are relatively simple, thus there are only a few problems that ever arise. The most common is incorrect split ratios. Either the split ratio has changed from previous values, or excessively low or high values are being used. If a change in peak size can not be attributed to any other source, check the split ratio value for changes. A higher split ratio than previously set accounts for smaller peaks while a lower split ratio accounts for larger peaks. A large change in the split ratio may also account for changes in relative peak sizes (discrimination). Make sure the changes can not attributed to other sources such as changes in the sample or detector.

If peaks are large and tailing, the split ratio may be too low. The low carrier gas flow rates with low split ratios severely reduces injector efficiency, hence giving tailing peaks. Also, check that the injector is not being used in the splitless mode. This would result in very large peaks and a large solvent front. The peaks shapes may range from good to very bad depending on the sample and column temperatures. Extremely small peaks that are difficult to see (especially if they are very narrow) may be an indicator that the split ratio is too high. Only a tiny amount of sample is reaching the column. Make sure a split ratio that is within the range for a given column diameter is being used.

7.4   Splitless Injectors
7.4.1 Description

A splitless injector does not split the sample, thus most of the vaporized sample is introduced into the column. This makes splitless injections suitable for trace level analyses, but not for high concentration analyses. The lowest concentration level depends on the sensitivity of the detector being used. The highest concentration level is primarily limited by the capacity of the column and sometimes the linear or mass range of the detector.

Most injectors operate in the split and splitless modes. Split only injectors exist, but they are usually only found in very low cost or speciality instruments. There is a heated injector, liner, incoming carrier gas line, split line and septum purge line. The split line has a solenoid which opens and closes this line to gas flow. After entering the injector, the carrier gas exits by way of the column, split line and septum purge line. The septum purge flow is very low and above the region where most of the sample vaporization occurs, thus very little sample is lost via the septum purge line.

Splitless injectors have two operational modes which occur in sequence during an analysis. At injection and for a short time afterwards, the solenoid closes the split line and there is no gas flow through the line (Figure 7-4a). This mode is usually called purge off. Upon sample injection and vaporization, the sample mixes with the carrier gas and is transported towards the column. Since the split line is closed at this time, the only place the sample can travel is into the column. Sample transfer into the column is very slow since the carrier gas flow in the injector is the same as the column flow (1-3 mL/min) during purge off. The resulting poor sample transfer is compensated by use of the solvent effect or cold trapping (Sections 7.4.3 and 7.4.4). After a preset time period (called the purge activation time or purge on time; Section 7.4.2), the split line is opened by the solenoid. The carrier gas flow that previously by-passed the injector during purge off now flows through the injector and out through the split line (Figure 7-4b). This mode is usually called purge on. The splitless injector now behaves just like a split injector. Any residual sample still in the injector is swept out of the split line (except for an inconsequential amount that enters the column). The inlet purge flow (not the septum purge flow) is usually set between 10 and 25 mL/min and should be set during the purge on cycle.

The inlet purge function is necessary to keep the solvent front from being excessively wide. The slow carrier gas flow rate in the injector during purge off requires an excessively long time to flush all of the sample from the injector and into the column. Since almost all of the sample is solvent, most of the residual sample in the injector is solvent. Failure to purge the injector of this solvent results in a very broad and tailing solvent front (Section 7.4.9).

Figure 7-4: Splitless Injector
Some details are omitted for clarity.   *The total carrier gas flow, flow
into the septum purge line and out the split line differs during purge off
for the two different injector designs (see Appendix F).

Too short purge activation times results in very small amounts of sample entering the column; too long purge activation times places more sample in the column, but creates a very large solvent front.

7.4.2 Selecting Purge Activation Times

The best purge activation time depends upon the carrier gas flow rate and the volatility of the sample compounds. In general, higher boiling point (lower volatility) compounds require longer purge activation times than lower boiling point compounds. More time is needed to thoroughly vaporize higher boiling point compounds and introduce them into the column in sufficient amounts. Lower boiling point compounds do not require as much time to thoroughly vaporize, thus short purge activation times are satisfactory.

As a guideline, purge on should occur when about 1.5 volumes of carrier gas has swept the injector. This is calculated by dividing the flow rate of the carrier gas into the injector liner volume (Equation 7-2). Most liners have a volume of 0.5-1 mL; most carrier gas flows are 1-3 mL/min. This results in typical purge activation times of 15-90 seconds. Experimenting with purge activation times slightly above and below the calculated time is recommended. The goal is to maximize analyte peak size and minimize solvent front size. Excessively long purge activation times increase solvent front size without an appreciable increase in peak size; too short purge activation times result in smaller than desired analyte peaks. For most applications, it has been found that longer purge activation times (40-90 seconds) are better than short purge times (20-40 seconds).

$$Sweep\ Rate = \frac{Liner\ Volume}{F_{column}}$$

F_{column} = Column flow rate (mL/min)
Liner Volume = $\pi\,r^2 L$ (cm)

Equation 7-2:   Injector Sweep Rate

7.4.3 Solvent Effect

Sample transfer into the column occurs during purge off (before the purge activation time). Sample transfer occurring over such a long time results in the vaporized sample being spread over a long length of column with a subsequent loss of efficiency. This problem is overcome by focussing the sample in a short band at the front of the column. The solvent effect is used to focus the sample in splitless injections.

The solvent effect depends on the condensation of the sample solvent at the front of the column. This is accomplished by maintaining the initial temperature of the column at least 10°C below the boiling point of the sample solvent. Since the column is below the boiling point of the solvent, it condenses in a thin film at the front of the column (Figure 7-5b). The solvent comprises the overwhelming majority of the sample and is usually the most volatile component present.   As the remaining sample components

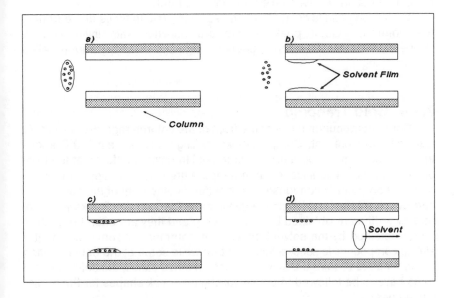

Figure 7-5: Solvent Effect
a) Injected sample   b) Solvent film forms upon condensation of the sample solvent   c) Compounds trapped in the solvent film   d) Solvent elutes from the column leaving the compounds behind in a narrow band

enter the column after the solvent, they are trapped in the solvent film (Figure 7-5c). This focusses the injected sample over a short length of column and overcomes the poor sample introduction problem. As the column is heated during the temperature program, the solvent volatilizes first and the remaining sample components are left behind in a short band (Figure 7-5d). As the column temperature continues to increase, the sample components begin to chromatograph as desired.

In some cases, better peak shapes are obtained when the initial column temperature is 20-30°C below the solvent boiling point. This is especially helpful when the first peaks elute close to the solvent front. If possible, use a sample solvent with a boiling point at least 25°C less than the most volatile sample component. The solvent needs to be the most volatile sample component. Any compounds more volatile than the solvent may enter the column before the solvent. Since the solvent film has not formed yet, these compounds are not trapped and focussed by the solvent film. The peaks that elute before the solvent peak are broad and distorted. If these peak are not of interest, their poor shapes are not important.

Some compounds do not need the solvent effect to focus at the front of the column. Good peak shapes are obtained with initial column temperatures above the sample solvent boiling point. This behavior is called cold trapping.

7.4.4 Cold Trapping

Some compounds focus at the front of the column regardless whether a solvent film is present. Compounds with boiling points at least 150°C above the initial column temperature undergo cold trapping and focus at the front of the column. The initial column temperature is low enough so that the sample compounds condense in a tight band at the front of the column - a narrow solvent film is no longer needed to focus the sample. Peaks with very narrow widths are obtained by cold trapping and they are often better than those obtained by the solvent effect. It is common for some of the later eluting compounds to successfully cold trap while some of the earlier eluting compounds to require the solvent effect. In these cases, the solvent effect rules have to be followed to obtain satisfactory peak shapes for the earlier eluting peaks.

7.4.5 Septum Purge for Splitless Injectors

The septum purge line can act as a miniature split line if the sample reaches the top of the injector during the vaporization and expansion process. The more volatile sample components are more likely to flash towards the top of the injector. Due to the lower sweep rate of splitless injectors, there is greater opportunity for the more volatile sample components to reach the top of the injector. The amount of sample loss varies depending on the design of the injector (i.e., forward vs. back pressure regulated; see Appendix F). The sample loss occurs during the purge off time (before the purge activation time) with most of the loss occurring within the first several seconds after injection.

Back pressure regulated injectors add the split line flow to the septum purge flow during the purge off time. This results in a very high gas flow (septum purge + split line) across the top of the injector during the time of maximum sample flashback. Potentially, large amounts of the most volatile sample components can be lost via the septum purge line. For this reason, the split line flow with back pressure regulated injectors used in the splitless mode should be kept fairly low (10-25 mL/min). Forward pressure regulated injectors maintain the same septum purge flow throughout the analysis. Sample loss via the septum purge line is much less for these types of injectors and often lower discrimination is evident also. Higher split line flows can be used to sweep forward pressure regulated injectors without extensive sample loss.

7.4.6 Splitless Injection Liners

Splitless injection liners are usually straight tubes with some having a restriction at one or both ends. The restrictions act as backflash barriers. They help to keep the sample inside of the liner during the expansion portion of sample vaporization. This helps to keep the injector lines and septum face much cleaner, which helps to minimize carryover or ghost peak problems. A bottom restriction acts as a guide to keep the column centered in the liner and away from the walls. Erratic peak shapes and sizes may be encountered if the column end is touching the inner wall of the liner. An upper restriction acts as a needle guide and helps to direct the syringe needle to same location for each injection. An upper restriction will require better alignment of autosamplers and precise manual injections to prevent damaging the tips of the needles.

The lowest cost liners are the straight tube liners. The cost increases if there are any restrictions. The double restriction liners are more difficult to clean since the openings are quite small. The most common splitless liner inner diameters are 2 and 4 mm. The 2 mm I.D. liners are more susceptible to backflash problems, but they provide slightly better efficiency. In most cases, the small gain in efficiency is not worth the increase in backflash (injector contamination) problems. Also, injection volumes are limited to ≤1 µL or significant backflash problems occur. Using glass wool in a splitless liner is not recommended. The glass wool causes excessive mixing and dilution of the vaporized sample with carrier gas. This may cause broader or smaller peaks. In cases where the injected sample contains a large amount of high boiling point or non-volatile compounds, the use of glass wool may be warranted. The slight losses in efficiency or peak size may be worth the benefit of the filtering effect of the glass wool. Thoroughly deactivated liners are more important with splitless injections. The sample spends a large amount of time in the liner, thus active compounds have greater opportunities to suffer from adsorption (peak tailing or size loss).

7.4.7 Column Position in Splitless Injectors

Each model of GC has a distance to which the column is inserted into the injector. This value is usually found in the column installation instructions for the GC. The distance recommended by the GC manufacturer is probably the best one for the widest variety of samples; however, it may not be the best one for a specific sample. Errors in the insertion distance, especially if the distance is too short, may result in severe problems. The column insertion distance primarily influences efficiency and discrimination. Sometimes, even several millimeters difference has a pronounced impact on these parameters. The best distance varies depending primarily on the nature of the sample compounds.

For most splitless applications, the manufacturer's recommended column position is usually near the bottom of the liner. If a bottom restriction is present, it is suggested that the column end be placed slightly above the opening of the restriction. It has been reported that there should be 1-1.5 cm between the end of the syringe needle and the column end. This requires the column to be inserted farther into the injector than normally recommended by the GC manufacturer. This column position seems to work best for samples containing very high boiling compounds (e.g., PAH's, steroids). Regardless of the type of liner, the position of the column in the liner may have a significant effect on efficiency and discrimination. Experimenting with column position, within the previously discussed guidelines, may result in the best efficiency or minimal discrimination for a particular sample. If optimizing these two parameters is not extremely important, using the GC manufacturer's recommended insertion distance will be satisfactory. Most importantly, consistent column position is critical if a high degree of reproducibility is required.

7.4.8 Other Aspects of Splitless Injectors

The use of a retention gap will often improve the peak shapes for compounds that rely on the solvent effect for focussing. The peaks closest to the solvent front are often slightly rounded or broad. A retention gap sometimes helps to sharpen these peaks with a potential increase in resolution.

Since splitless injections deposit a large amount of sample in the column, a retention gap acts as a guard column. All of the sample is focussed in retention gap instead of the analytical column. Non-volatile residues and compounds that can damage the stationary phase are not concentrated on the stationary phase. Keeping these types of materials away from the stationary phase improves column performance and lifetime.

7.4.9 Common Problems with Splitless Injectors

One of the most common mistakes made with splitless injections is using an initial column temperature that is too high. The solvent effect does not occur if the sample solvent does not condense in the front of the column. The peaks are broad, rounded or distorted with the earlier eluting ones being the worst; if any of the sample compounds cold trap, those peaks are well formed (Figure 7-6).

Another problem area involves the purge on/off function. If the purge on function does not activate (i.e., it should turn on after 10-90 seconds) or occurs after a very long time, a very broad and tailing solvent front is obtained (Figure 7-7). If the purge is on during the injection (i.e., it should be off), the solvent front and solute peaks are much smaller than expected (Figure 7-8). Most of the sample escapes from the split vent since it was open. Essentially, a split injection is made.

If a peak shape problem occurs that can not be attributed to a solvent effect problem, a polarity mismatch may be the cause. This usually occurs when the solutes and solvent are similar in polarity while the stationary phase is different. Instead of condensing into an even, thin film at the front of the column, the solvent beads up or smears. The solutes are dispersed into the uneven film, thus they are not focused into a narrow band and the peak shapes are poor. Only compounds that are very soluble in the solvent are affected by the polarity mismatch. For example, only the polar alcohols in a mixture of solvents in water exhibit peak shape problems when using a relatively nonpolar column; the immediate and nonpolar solvent peaks are well formed. The polar alcohols are very soluble in the very polar water while the other solvents are much less soluble. Changing the solvent to one that

Figure 7-6:   Solvent Effect Violation
            Column: DB-5, 15 m x 0.25 mm I.D., 0.25 µm
            Oven: 90°C for 1 min, 90-225°C at 15°/min
            Injector: Splitless, 30 sec purge on time
            Sample: Hydrocarbons in n-hexane (B.P. 69°C)

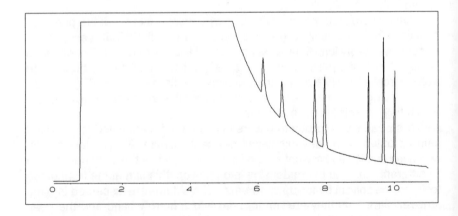

Figure 7-7:   Lack of a Purge Function
            Column: DB-5, 15 m x 0.25 mm I.D., 0.25 µm
            Oven: 50°C for 1 min, 50-150°C at 10°/min
            Injector: Splitless, No purge on time
            Sample: Column test mixture in n-hexane

Figure 7-8: Purge On Error for a Splitless Injection
Note the smaller size of the peaks and the solvent front
Column: DB-5, 15 m x 0.25 mm I.D., 0.25 μm
Oven: 50°C for 1 min, 50-150°C at 10°/min
Injector: Splitless
Sample: Column test mixture in n-hexane

is closer in polarity to the stationary phase improves the peak shapes. If the sample solvent can not be changed, the use of a retention gap sometimes improves the peak shapes.

A peak shape problem with a low concentration compound eluting close to a high concentration compound is another problem encountered in splitless injections. Instead of partitioning only into the stationary phase and carrier gas, the low concentration compound also partitions into the band of high concentration compound in its vicinity.   As the band of high concentration compound moves through the column, it may pull or push a portion of the low concentration compound along with it. This often causes a tail or shoulder on the side of the peak closest to the high concentration compound. The amount of peak distortion is dependent on the concentration of the high level compound. A shift in the low level compound's retention time may also be observed as the concentration of the high level compound changes. High and low concentration sample compounds with similar polarities exhibit the greatest peak shape problems; compounds with large polarity differences have slight or no peak shape problems.

Injection size has a direct effect on the length of the solvent film. The length of the solvent film increases with increased injection volumes. Small losses in efficiency may be observed as injection sizes start to exceed 2 µL. The solvent film length is shorter for larger diameter columns since the surface area per unit length of column is greater. In some cases, the reduced solvent film length may offset the decreased efficiency of a larger diameter columns. This effect primarily occurs for the peaks that elute closest to the solvent front. As peak retention increases, the benefits of the larger diameter column diminishes until no gains are obtained.

7.5   Megabore Direct Injectors
7.5.1 Description

A Megabore injector introduces nearly all of the vaporized sample into the column. This injector is limited to use with Megabore (0.53 mm I.D.) columns. Megabore columns were intended to replace packed columns while still using most of the original packed column injector. Megabore direct injectors are also available as an option on most models of GC's. A packed column injector can be made compatible with a Megabore capillary column with use of a simple conversion kit. These kits are available from the GC and capillary column manufacturers.

Megabore direct injectors are quite easy to use and suffer from relatively few problems. Samples over a wide concentration range are suitable for Megabore columns. Thus trace to high concentration analyses are possible with Megabore direct injectors. Like all injectors, the lowest concentration level depends on the sensitivity of the detector being used; the highest concentration level is primarily limited by the capacity of the column and, sometimes, the linear or mass range of the detector.

Megabore direct injectors consist of a heated injector, liner, incoming carrier gas line and, sometimes, a septum purge line (Figure 7-9). After entering the injector, the carrier gas exits by way of the column or the septum purge line (if present). The septum purge flow is very low and above the region where most of the sample vaporization occurs; thus very little sample is lost via the septum purge line.

After injection and vaporization, the sample vapors are mixed with the carrier gas. The carrier gas travels into the column, thus transporting the sample into the column. Since there are no other outlets for the carrier gas, nearly all of the vaporized sample enters the column. The sample is swept into the column at a moderately high rate.

Typical carrier gas (helium or hydrogen) flow rates with Megabore columns are 4-10 mL/min. This results in sweep rates of 10-100 seconds for most Megabore direct injectors. Smaller diameter columns have flow rates of 1-3 mL/min. These flow rates result in sweep rates of 30-400 seconds. Such long sweep times result in very broad peaks especially the solvent front. This makes columns with diameters of 0.32 mm or less incompatible with Megabore direct injectors.

Figure 7-9:   Megabore Direct Injector
Septum purge line may not be present on some Megabore direct injectors especially those converted from packed column injectors.

Recently, columns with 0.45 mm I.D. have been introduced. These columns are compatible with Megabore direct injectors and liners. They use the same hardware and ferrules, and are interchangeable with 0.53 mm I.D. columns in nearly every application.

7.5.2 Megabore Direct Injection Liners

The main difference between different Megabore direct injectors is the liner. The type of liner has a major impact on the quality of the chromatography obtained with a Megabore column. The primary liner styles are the straight tube and the direct flash.

The most common and lowest cost liner is the straight tube (Figure 7-10a). As the vaporized sample is carried towards the column, some of it can travel past the column. Also, any sample that backflashes can travel out of the top of a straight tube liner. These vapors now have to be swept by the carrier gas from these areas and back towards the column. This requires extra time and results in severe tailing of the solvent front. The severity of the solvent front tail is more pronounced for large injection volumes, volatile sample solvents, low carrier gas flows and very hot injectors.

The other type of Megabore direct injector liner is the direct flash vaporization (Figure 7-10b). It is distinguished by the restriction at the top and another restriction several centimeters further down. The exact position of the bottom restriction varies depending on the source of the liner. The Megabore column is pushed up into the liner until it can not move any further. The columns fit into the tapered region of the bottom restriction and forms a leak-free seal. The polyimide column coating is compressible and forms a tight seal between the column and the liner. This prevents any vaporized sample from flowing around or past the column. The restriction acts to funnel all of the sample into the column. The outer diameters of non-Megabore (<0.45 mm I.D.) columns are too small and these columns pass through the lower restriction. The column and liner do not form a tight seal, thus defeating the purpose of this type of liner. The upper restriction acts both as a needle guide, and more importantly, as a backflash barrier. If any sample flashes to the top of the liner, it is prevented from leaving by the upper restriction. Containing the vaporized sample within the liner chamber results in minimal backflash and a fast sweeping of the sample into the column. This results in better peak efficiencies, and more dramatically, reduces the width and amount of tailing of the solvent front. This improvement is most noticeable when compared to a straight tube liner. A direct flash vaporization liner is recommended for Megabore direct injectors,

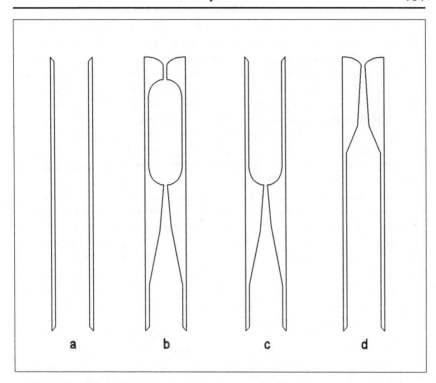

Figure 7-10: Megabore Direct Liners
a) Straight tube   b) Direct flash   c) Direct flash without upper restriction   d) Hot on-column

especially for injection volumes above 2 µL or low boiling sample solvents.

There are two variations on the direct flash vaporization liner. One has the upper restriction is removed so that the liner can be easily cleaned or packed with silylated glass wool (Figure 7-10c). The benefits of the backflash barrier created by the upper restriction is no longer present, but the ease of cleaning or use of glass wool may outweigh the loss.

The other type of direct injection liner is called a hot on-column liner (Figure 7-10d). The restriction where the column forms the seal with the liner is at the top of the liner. A standard 26 gauge syringe needle fits inside a 0.53 mm I.D. column. The injection is made directly into the column, thus bypassing the liner. In some ways, this is not a vaporization technique. The injector is usually maintained at an elevated temperature, but at one lower than normally used for vaporization techniques. Injector temperatures of 100-200°C are recommended with the lower range more commonly used with low boiling point solvents. Excessively high temperatures cause the sample solvent to significantly expand within the column. This results in severe peak and solvent front broadening or backflash problems. Since it is easy to flood the column with sample, the maximum sample volume for hot on-column liners is 0.5 µL. Injecting larger sample volumes results in distorted peak shapes and large solvent fronts. Even with small injection volumes, broad solvent front and peaks are common with hot on-column injection liners. The advantage of using a hot on-column liner is minimal discrimination and decreased temperature induced sample decomposition. Improved peak shapes are often obtained if a retention gap is used with a hot on-column injector liner. The added benefit of longer column lifetime is usually realized. The sample is directly deposited in the column, thus exposing it to a large amount of non-volatile materials. These can interfere with proper chromatography and reduce column life. The retention gap helps to minimize the problems associated with the buildup of non-volatile materials at the front of a capillary column.

The type of liner used with a Megabore direct injector may be limited by the design or source of the injector. Some require specific liners as designated by the manufacturer of the GC or conversion kit. Liner availability and cost should be considered before deciding whether to purchase a pre-configured Megabore direct injector or a packed column injector and a conversion kit from another source. The conversion kits are easy to install and use, and are usually available at relatively low cost.

7.5.3 Septum Purge for Megabore Direct Injectors

Some Megabore direct injectors do not have a septum purge function. This is most common for packed column injectors converted to a Megabore direct injector. The septum purge line can act as a miniature split line if the sample reaches the top of the injector during the vaporization and expansion process. The more volatile sample components are more likely to flash towards the top of the injector. This will be a greater problem for the liners without some type of upper restriction. Losses of volatile sample compounds are minimized with a liner having an upper restriction. Septum purge flows of 0.5-5 mL/min are still recommended for Megabore direct injectors. Lower flow rates are best for the open top liners.

7.5.4 Column Position in Megabore Direct Injectors

For the straight tube type of direct injection liner, the column should be located near the bottom of the liner. If the column opening is above the bottom portion of the liner, the carrier gas has to sweep the liner volume many times before all of the sample vapors are transferred into the column. The solvent peak (and perhaps the solute peaks) become very wide and have an extreme tail if the column is positioned too high in the liner.

Improper installation of the column in a direct flash vaporization liner results in a broad and tailing solvent front. The shape of the peaks may also suffer. The correct sealing of the column in the liner restriction is critical. Any leakage of carrier gas around the seal will result in a broad and tailing solvent front. The worse the leak, the worse the solvent front. A column that has not been inserted far enough into the liner does not seal, but leaks. A column that is pushed into the restriction with excessive force may chip or break; it will not seal properly but will leak instead. Leaks also occur if the column end is poorly cut. The column end needs to be even, without any burrs or chips, and at a 90° angle to the column wall. Also important is the cleanliness of the sealing surface of the liner. Any debris in the restriction prevents the proper sealing of the column. Small pieces of septa may fall into the restriction and become lodged when a column is removed. On occasion, a small portion of column or polyimide coating adheres to the walls of the liner restriction upon column removal. This prevents proper sealing. Repeated column sealing difficulties warrant the removal of the liner for inspection since its interior is not visible during installation.

7.5.5 Other Aspects of Megabore Direct Injectors

The low efficiency of Megabore columns and injectors make closely eluting peaks more difficult to resolve. One method of potentially improving injector efficiency is to use a retention gap. The retention gap helps to focus the sample in the front of the column. Any focussing of the sample helps to decrease peak widths, and thus improves resolution. The retention gap also acts as a guard column. This is beneficial since Megabore direct injections place most of the sample into the column. Non-volatile residues and harmful compounds are kept away from the analytical column. Improved column life and performance and less column maintenance may result.

Using an initial column temperature at least 10°C below the boiling point of the sample solvent may help to focus the injected sample. The solvent effect (Section 7.4.3) occurs and this often improves peak widths. The greatest improvement occurs for peaks that elute closest to the solvent front.

7.5.6 Common Problems with Megabore Direct Injectors

Probably the most common problems with Megabore direct injectors involve the position of the column in the injector and contamination. For straight tube liners, having the column too low in the liner causes peak tailing problems. Having the column too high in the liner is usually not serious and just changes the amount of discrimination. For the direct flash vaporization liners, a leak around the seal area is the biggest problem. Very large and tailing solvent fronts are seen with a poorly sealed column.

Besides the septum purge and the column, there are no other flow paths leaving the injector. The entire sample is introduced into the injector. If backflash occurs, the septum purge and incoming carrier gas lines are the only outlets for the vapors. Contamination of the septum purge line is not a major problem since the gas is flowing out of the injector. Contamination of the incoming carrier gas line with backflash residues is a problem since the materials can be carried back into the injector and into the column at a later time. This leads to ghost peaks or baseline interferences.

7.6   Specialized Vaporization Injectors

There are several recent options or modifications to vaporization injectors that may improve injector and column efficiency, simplify carrier gas velocity and flow setting, or reduce analysis times. Depending on the sample and column, some or all of these benefits can be obtained. These injectors are not available from every GC manufacturer and may be limited to only one or two sources. While the details about these injectors are beyond the scope of this book, the basics are presented. Details about the specific injectors can be obtained from the various manufacturers.

7.6.1 Pressure Programmable Injectors

As the name implies, a pressure programmable injector can adjust the carrier gas pressure during a GC run. This can be desirable due to some of the properties of gases. Carrier gases become more viscous at higher temperatures. At a fixed pressure, the carrier gas velocity decreases as the gas becomes more viscous. Standard vaporization injectors maintain a constant carrier gas pressure throughout the run, thus carrier gas velocity decreases during a temperature program. While this is not a problem, there are advantages to being able to adjust or maintain a constant carrier gas linear velocity or flow during a run.

The column length, diameter, carrier gas identity, and desired average linear velocity is entered from the keyboard or software, and the temperature program is obtained by the GC from its memory or software. The system calculates the carrier gas pressure needed to obtain the desired velocity or flow at appropriate times and temperatures. The pressure is adjusted by the injector system during the run to obtain the desired velocity or flow results.

Pressure programmable injectors have a built in program to increase the carrier gas pressure to maintain a constant linear velocity throughout the run. The most obvious impact of constant velocity is to decrease retention times especially for the later eluting compounds. Shorter run times are realized. Depending on the column, carrier gas and its linear velocity, improved resolution may be obtained for some samples at a constant linear velocity. It is somewhat difficult to predict whether any resolution improvements will occur, thus some trial and error is usually involved. Another advantage of constant linear velocity or flow situations is related to the behavior of some detectors. Some detectors are concentration dependent and their response is proportional to the volume of carrier gas flowing through them. As the carrier gas flow changes during a temperature program run, the chromatogram baseline drifts proportionally to the change in the carrier gas flow. If the carrier gas flow is kept constant, this baseline drift will not occur. Carrier gas related baseline drift problems are the most noticeable with NPD, TCD and GC/MS systems.

Pressure programmable injectors can be instructed to change the carrier gas pressure to set values at specific times. This can be used to adjust the linear velocity to its maximum efficiency value during any part of a run. Finding these optimum values involves trial and error, thus requiring a large time investment. The efficiency gains may not be worth the development time and the resulting complexity of the carrier gas parameters. Pressure programming is mainly beneficial in optimizing the injector efficiency. In general, high carrier gas flow rates in the injector at the time of injection results in better peak widths. Some injectors use very slow carrier gas flow rates. Increasing the carrier gas flow to make the injector more efficient results in a large loss of column efficiency. A programmable injector can be programmed to use a high flow rate during the injection then quickly drop down to one that is best for the column. Both the injector and column efficiency can be optimized without sacrificing one for the other. Split injectors can benefit from pressure programming also. Very low split ratios can not normally be used due to the excessively low carrier gas flow rates in the injector. If the carrier gas flow is increased for only a few seconds at the start of the run, very low split ratios can be used. The momentary high flow rate in the injector increases the injection efficiency so that low split ratios can be tolerated.

Probably the most overlooked benefit of pressure programmable injectors is the ease with which linear velocities and split line flows can be set. This is done via the GC keyboard or software. As long as the correct carrier gas, column length and diameter is supplied to the system, the linear velocity is properly set for the given column temperature. This eliminates some of the potential errors that occur with manual setting of the linear velocity and split ratio.

7.6.2   Temperature Programmable Injectors

Injectors are usually maintained at a fixed temperature during an analysis. As long as the injector temperature is hot enough to rapidly vaporize the sample, there are no significant problems. Temperature programmable injectors can very rapidly heat an injector during the GC run. This creates a number of interesting possibilities.

If there is a large enough temperature difference between the sample solvent and most volatile sample compound, elimination of the solvent front is possible. An on-column injection is used to deposit the sample into the column. The injector is then heated to a temperature where the solvent elutes from the column while the compounds remain focussed at the front of the column. After a fixed length of time, the injector temperature is raised and the compounds move into the part of the column in the GC oven for separation. If the column is very retentive for the compounds and the solvent elutes at a low temperature, multiple injections can be made on top of each other. After the first injection's solvent has eluted, another injection is made. The solvent is eluted, but all of the highly retained compounds remain focussed at the front of the column. Very dilute samples can be concentrated in the column so that they can be easily detected.

Some compounds degrade upon sudden changes in temperature. The sudden change from room temperature to hot injector temperature can increase the amount of degradation. Sometimes less breakdown occurs if the injector temperature is maintained at a lower temperature at the time of injection then the temperature is raised to vaporize the sample. A temperature programmable injector allows the injection to be made at a lower temperature followed by a rapid increase in the injector temperature. This exposes the sample to a less severe temperature change upon injection. Temperature programmable injectors may be helpful in cases where samples contain a semi-volatile materials that contaminate the injector. These materials eventually enter the column and create baseline disturbances and ghost peaks. By heating the injector to a very high temperature while the column is held at a high temperature at the end of each run may help to clean out the injector. The contaminants are volatilized at the higher injector temperature and quickly elute from the column held at the high temperature. This is rough on the column, but may be worth the avoidance of contamination problems.

7.6.3 Septumless Injectors

Septumless injectors of various designs have been available for many years. Most are available as a simple addition to many designs of existing vaporization injectors. By eliminating the septum, problems attending septum changes and the possibility of leak and bleed problems are avoided. Most septumless injectors use a duckbill (two ribbons of metal with a flared opening resembling a duckbill) or a pair of offset spheres to seal the injector. Springs or carrier gas pressure is used to keep the two surfaces together and leak free. As the syringe needle is pushed into the injector, the two surfaces are forced part. The needle can now enter the interior of the injector for sample introduction. The two surfaces close and reseal the injector when the syringe needle is withdrawn.

The types of samples analyzed, carrier gas pressures and injection technique all influence the performance of septumless injectors. Mixed results have been reported, and there seems to be no consistent pattern related to the successful use of such injectors. The syringe needle needs to be free of rough areas, nicks or burrs. These scratch the surface of the seals and cause them to leak after a small number of injections. Periodic replacement of the seals is also necessary.

7.7   Cool On-Column Injectors
7.7.1 Description

On-column injection eliminates the vaporization processes as the mechanism for sample introduction into the column. On-column injectors deposit a liquid sample directly into the column without going through a vaporization step. On-column injection eliminates many of the problems associated with vaporization injectors. Eliminating sample vaporization eliminates backflash problems. Since the sample is not vaporized and subject to expansion, it does not flow out of the injector. On-column injection does not subject the sample to high temperatures which makes it ideal for heat sensitive compounds. On-column injection places all of the sample in the column. Discrimination problems are avoided since the amount of each compound entering the column is dependent on its concentration and not its volatility. On-column injection makes possible the analysis of some high boiling compounds by capillary GC that are not possible using a vaporization injector. Many high boiling point compounds do not vaporize fast enough in vaporization injectors to permit their rapid introduction into the column. This is especially true for the fast injection modes such as split. Since the entire sample is placed directly in the column, the high boiling point compounds have the opportunity to chromatograph and separate. The characteristics of cool on-column injections make it ideal for several types of compounds or samples - very high boiling point compounds, heat sensitive compounds and samples that require minimal discrimination effects.

A dedicated on-column injector is needed to make on-column injections (Figure 7-11). The primary purpose of the injector is to guide a on-column syringe needle into the column. The syringe has a fine needle that fits into the column. Depending on the injector, columns with inner diameters as small as 0.25 mm can be used. Carrier gas enters the injector and is directed into the column. A stopcock or valve is used to keep the carrier gas from leaking from the needle entrance. The stopcock or valve is opened so that the syringe needle passes through and enters the column. A small amount of carrier gas may escape from around the needle, but this is normal and does not cause any problems. Usually there is some type of insert that helps to guide the syringe needle into the column and to keep the carrier gas from escaping around the column. The injector is usually kept at ambient temperature since immediate sample vaporization is not needed or desired. Within the injector body, a short section of column is cooled by forcing ambient air around the column using the secondary cooling option (Section 7.7.3). This cool zone at the front of the column is needed to obtain good peak shapes. By only cooling a small section of column outside the GC oven, the entire oven does not need to be maintained at a low temperature.

7.7.2   Solvent Effect and Cold Trapping for Cool On-Column Injections

A 1-2 µL injection of a liquid can flood a large region of the column. This is a greater problem with smaller diameter columns. The large flooded zone results in a broad sample band at the front of the column which usually causes broad peaks. The sample needs to be focussed into a short band to prevent poor peak shapes. The solvent effect or cold trapping is used to focus the sample in cool on-column injections.

Figure 7-11: Cool On-Column Injectors
a) Grob Cool On-column Injector
b) Duckbill On-column Injector

Utilizing the solvent effect requires the column to be at least 10°C below the boiling point of the sample solvent. The solvent condenses at the front of the column in a short band since the column temperature is below the boiling point of the solvent. This film of solvent traps the rest of the sample and focuses it in a short band. Any sample component more volatile than the solvent enters the column prior or at the same time as the solvent. The peaks representing those compounds may be broad and distorted to varying degrees. If those peaks are not of interest, this is not a problem.

In some cases, cold trapping can be used to focus the sample. Cold trapping occurs when the column temperature is at least 150°C below the boiling point of a compound. Any sample compounds that fulfill this condition focus in a short band at the front of the column. These compounds have good peak shapes without the aid of the solvent effect. To obtain good results with a cool on-column injector, either the solvent effect or cold trapping needs to occur. After the sample has entered and focussed in the column, the column temperature is raised to facilitate separation. A temperature program is almost always used to start the column at a lower temperature and to raise the temperature higher enough to elute all of the compounds within a short time period.

7.7.3 Secondary Cooling

Since the solvent effect or cold trapping is necessary to obtain good peak shapes, the initial value of the temperature program is usually relatively low. This temperature is often much lower than that needed to adequately separate the sample compounds. Most cool on-column injectors have a secondary cooling function that eliminates the necessity of cooling the entire column down to a low temperature.

A small fan is used to force room temperature air around a small section at the front of the column (Figure 7-11). The use of secondary cooling means that the initial column (oven) temperature does not have to be below the boiling point of the sample solvent. After injection of the sample, the secondary cooling is discontinued and the column rapidly assumes the oven temperature. Temperature programming can then commence to separate the compounds.

Secondary cooling may be accomplished without some type of forced air circulation. One on-column injector design uses a telescoping injector unit that pulls a small section of column outside of the column oven. The sample is injected while the column section is outside of the oven and at room temperature. This has the same effect as forced air secondary cooling. After the injection, the exposed section of column is carefully pushed back into the heated column oven to start the chromatographic process.

The main advantage to secondary cooling is the reduction of time between analyses. Shorter analysis times are possible since the temperature program can be started at a higher initial temperature. Also, less time is needed to cool down the oven between runs. One disadvantage of secondary cooling is the need for a retention gap.

7.7.4 Retention Gaps and Cool On-Column Injectors

Whenever secondary cooling is used, a retention gap is needed to maintain good peak shapes. Any time a sample band moves from a region of high retention to a region of low retention, band broadening occurs. Band broadening creates broad peaks. This occurs for cool on-column injections with secondary cooling when a retention gap is not used. The sample band moves from a region of higher retention (e.g., the portion of the column cooled by the forced air) to one of lower retention (e.g., the portion of the column in the GC oven). The use of a retention gap corrects this problem.

A 1-5 meter piece of deactivated fused silica tubing is an adequate retention gap. Tubing of equal or slightly larger diameter is often best for the retention gap. Since there is no stationary phase in the retention gap, there is no compound retention in this section. When a retention gap is installed, the portion of column in the on-column injector has no stationary phase. The sample now travels from a region of no retention (i.e., the retention gap) into one of much higher retention (i.e., column containing stationary phase). This occurs even though the portion of the column containing stationary phase is at a higher temperature than the front section of the retention gap. Failure to use a retention gap for cool on-column injections with secondary cooling usually results in very broad peaks. If secondary cooling is not used, a retention gap is not necessary, but still recommended. It will usually improve the peak shapes and help to protect the column from non-volatile sample components.

7.7.5 Other Aspects of Cool On-Column Injection

The entire sample is deposited in the column with on-column injection. The biggest problem is with the non-volatile components in the sample. These non-volatile materials rapidly accumulate in the column and create chromatographic problems. In general, column contamination is a greater problem with on-column than vaporization injectors. The use of a retention gap, in addition to providing better peak shapes, helps to keep the non-volatile materials away from the column. This reduces the frequency of problems related to non-volatile residue deposition and may result in increased column lifetime.

Cool on-column injection volumes are limited to 0.5-2 µL. Substantial peak broadening or distortion may occur upon injections greater than 2 µL. The large volume of solvent spreads over a large length of column. Even though the film may be dispersed uniformly, the excessive solvent film length causes the broad peaks.

If a retention gap is not used, stationary phase-solvent polarity mismatch may be a problem. If the solvent is of significantly different polarity than the stationary phase, distorted peak shapes may occur. The problem is compounded if the solutes are similar in polarity to the solvent. The use of a retention gap may improve peak shapes when a polarity mismatch occurs.

7.7.6 Common Problems With On-Column Injectors

The most common problem with on-column injections is the rapid fouling of the column. Problems with peak shape and size, separation and baseline disturbances are the most common symptoms of a contaminated column. The use of a retention gap (guard column) is recommended to minimize the onset of column contamination problems. The retention gap will have to be trimmed or changed as the non-volatile materials accumulate.

If peak shape problems can not be attributed to column contamination, the lack of a retention gap or the failure of the sample to respond to the solvent effect or to cold trapping are possible causes. Anything that prevents focussing of the sample or causes the sample band to spread or smear will create broad peaks. A poorly installed retention gap union may cause peak tailing. The tailing is usually worse for the more volatile sample compounds.

Mixed sample solvents may cause broad or split peaks. This usually occurs when the sample compounds are significantly more soluble in one of the solvents. The problem becomes worse as the difference in the boiling points of the two solvents increases. Little can be done to correct this problem besides removing or evaporating one of the solvents.

7.8   Injection Techniques

7.8.1 Syringe Filling Techniques

It may seem trivial, but the method used to fill the syringe with sample can have a surprising impact on the results. Erratic peak sizes are the most common problem with poor syringe technique. Peak shape problems can also occur, with spilt or fronting peaks being the most common.

It is important that the syringe needle does not contain any sample. When the needle is inserted into the hot injector, the needle is heated very rapidly. Volatile compounds in the needle, including the solvent, may begin to volatilize and leave the needle before the syringe plunger is depressed. This introduces a small amount of sample into the column before the rest. This usually creates a small peak immediately in front of each peak; in some instances, a tail on the front of the peak results. The separation between the two peaks or the severity of the front side tail increases as peak retention becomes longer.

The markings on the syringe barrel rarely account for the volume of the needle.   When measuring sample volume, this factor needs to be considered. Most needles on 10 µL syringes have a volume of 0.8-1µL. If 1 µL of liquid is measured in the syringe barrel, the actual volume is about

2 µL. A 2 µL volume as measured in the syringe barrel is actually about 3 µL which is not double the measured 1 µL volume.

Accurate and precise injections requires more than just pulling a volume of sample into the syringe. There are two techniques that aid in making accurate volume injections. The air plug method involves pulling 1-2 µL of air into the syringe before the sample is pulled into the syringe. Additional air is pulled into the syringe following the sample to ensure that no sample is in the needle (Figure 7-12a). The first plug of air helps to push all of the sample out the syringe needle and into the injector. Care needs to be taken to make sure that any residue of previous samples or solvents are not in the syringe. Removing traces of solvent from the syringe can be difficult and time consuming. Another syringe technique that eliminates this problem is the solvent plug method. This method first involves pulling about 1 µL of clean solvent into the syringe. Ideally, the solvent should be the same as the sample solvent. This clean solvent is followed by a 0.5-1 µL plug of air then the sample. Additional air is pulled into the syringe following the sample to ensure that no sample is in the needle (Figure 7-12b). The solvent and air helps to push all of the sample out the syringe needle and into the injector. The solvent also helps to clean the syringe needle. The extra solvent does increase the size of the solvent front, but this is usually not a problem.

7.8.2 Injection Speed

The different injection modes require different syringe techniques when making manual injections. The difference is primarily the speed of injection. Injection speed can be thought of as the speed of plunger depression and insertion/withdrawal of the needle from the injector. In general, relative injection speeds follow the order of split $\geq$ on-column > Megabore direct > splitless.

The high carrier gas flows in the split injector rapidly sweeps the vaporized sample into the column. For this reason, the entire sample needs to be introduced into the injector very quickly. A fast injection is needed to obtain the best efficiency. The entire process of needle insertion, plunger depression and needle withdrawal should take 1-3 seconds. Slow plunger depression may cause slightly broadened peaks or, in some severe instances, split peaks. The plunger depression should also be uniform in speed and the same from injection to injection. As soon as the plunger is fully depressed, the syringe should be immediately withdrawn from the injector. Slightly broader or tailing peaks can result if the syringe remains in the injector too long.

Figure 7-12:   Syringe Filling Techniques

The slower carrier gas flow rates in Megabore direct injectors allow a slightly slower injection speeds than with split injectors. Even though slower rates are satisfactory, it is recommended to make injections in the 1-3 second range. Slower injection speeds, unless very slow (e.g., >5 seconds), does not have a noticeable effect on the peak shapes.

Unlike split injections, the sample is swept into the column very slowly for splitless injections. Rapid sample introduction into the injector is not nearly as critical since it can take several seconds to transfer most of the sample into the column. It has been reported that syringe plunger depression at about 1 second per µL of sample is best for splitless injections. Faster injection rates do not seem to have significant detrimental effects.

The injection rate for on-column injections needs to be fast. With slow injections, a portion of the sample may be lost because it adheres to the inner walls or tip of the syringe needle. Fast injections help to spray the sample into the column away from the needle, thus avoiding partial sample introduction. Using a solvent plug helps to displace the entire sample out of the needle; however, the added solvent may slightly increase peak broadening.

A change in injection speed may be warranted when analyzing samples with high boiling point compounds. With any injection without a solvent plug, there is a small portion of sample that remains behind in the syringe needle after plunger depression. With vaporization injectors, the syringe needle rapidly heats and volatilizes some of the more volatile compounds remaining in the needle. These can now enter the injector, and thus also the column. Some of the higher boiling compounds may still remain behind on the walls of the needle. This portion of the sample is not introduced into the injector and contributes to the loss of high boiling point compounds. This type of compound loss is called syringe discrimination. It is more severe with fast injection speeds such as those used for split injections. The use of the hot needle technique may help to decrease syringe discrimination. This technique requires that there is no sample, only air, in the syringe needle. The needle is inserted into the injector and plunger depression occurs 1-2 seconds afterwards. This allows thorough heating of the needle by the hot injector. This gives the higher boiling point (i.e., slower vaporizing) compounds a better opportunity to vaporize and be introduced into the injector. It is important to have no sample in the needle or a small portion is prematurely introduced into the column. This normally results in split peaks with the first one of the pair being very small. More frequent septum changes may be required when using the hot needle injection technique. A better seal around the needle is needed when it is left in the septum for longer times.

7.9   Autosamplers

Autosamplers are primarily used to allow unattended sample injections. Another benefit is the greater reproducibility of autosampler injections over manual injections. Injection speed and sample amounts are controlled very precisely, resulting in peak size reproducibilities superior to manual injections. With these benefits comes the need to consider other factors usually not present with manual injections.

Depending on the model of autosampler, the speed of sample injection can be controlled to different extent. Most autosamplers, especially the newer models, perform very fast injections. Fortunately, the speed can usually be controlled by selecting different injection speed settings. The sample, injector and liner dictate which injection speed is the best. Experimenting with several of the possible speed settings may be worth the effort. Improved peak shape or size, better reproducibility or less discrimination may be obtained at a particular injection speed.

There are usually only a few autosampler options for syringe loading methods. The ones available depend on the model of the autosampler. Whether a solvent or air plug can be used, whether there is sample in the needle and available sample volumes are controlled by the flexibility of the particular autosampler.

If syringe solvent wash steps are available with an autosampler, they should be used. This prevents residual amounts of the previous sample from contaminating the next sample. This is a particular problem when a low concentration sample follows a high level one. Even small traces of a high concentration sample are significant when compared with a lower level sample. One of the wash solvents should be the sample solvent. This solvent is usually the most effective in removing previous sample materials. If more than one solvent is used for syringe washing, make sure that each successive solvent is miscible. For example, do not use methanol and hexane in succession; they are immiscible and this hinders the cleaning of the syringe. The solvent wash containers should be frequently cleaned. Traces of the numerous previous samples can accumulate in these vials upon evaporation of the wash solvent. If the vials are only refilled with clean solvent, the trace amounts of contaminants slowly build up to higher levels. After prolonged use of the vial, these compounds can build up to levels where they can contaminate the syringe instead of cleaning it.

When using volatile sample solvents or analyzing volatile compounds extra care needs to be taken to completely seal the autosampler vials. As the samples sit in the autosampler tray, the more volatile sample components can evaporate from a poorly sealed vial. If the sample solvent evaporates, the sample compounds becomes more concentrated. The peaks are larger than they should be, thus higher concentrations than the actual values are reported. If some of the more volatile sample compounds escape from the vial, a reduction in the peak size for these compounds occur. The more volatile the compound, the greater the loss from the vial. Also, there is a time factor that has to be considered. The longer the vial remains in the autosampler tray before injection, the greater the loss of solvent or compound. The error for the first sample injection is negligible while the error for the sample injected after 10 hours in the autosampler tray is more severe. The room temperature also has an effect on the amount of sample evaporation. Greater loss occurs when the room temperature is higher. Variations in room temperature throughout a day can alter the influence of room temperature on sample loss from the vial. The combination of sample solvent loss, volatile compound loss, time in the vial, and room temperature variations can make peak size irreproducibililty problems extremely variable. By using good quality vials and caps and properly sealing the vials, these types of problems can be avoided.

Larger diameters (i.e., 22 gauge) syringe needles are often used with autosamplers. This helps to prevent bending of the needles by the autosampler which is more common with the smaller diameter needles (i.e., 26 gauge). Also, the larger needles are less prone to become plugged with particulate matter such as pieces of septa. Since autosamplers are often left unattended, a plugged needle is not detected until the chromatograms are examined. This may result in many wasted hours and the need to reinject the samples. The larger needle pierce large holes in the septum. This usually necessitates more frequent septum changes. Even though the sample tray holds 100 vials, only 50 can be run during one session due to the need to change the septum after 50 injections. The use of a septumless injector may solve this problem providing this style of injector is compatible with the samples.

7.10   Injector Septa
7.10.1   Introduction
The septum's primary function is to seal the injector, thus preventing vaporized sample and carrier gas from leaving and air from entering the injector at the point of syringe introduction. Since a syringe needle repeatedly punctures the septum, it has to be frequently changed to prevent leaks. A septum leak may result in retention times and peak size variations which makes sample analysis more difficult and inaccurate. More serious is the leakage of air into the injector which exposes the column to high concentrations of oxygen. Oxygen accelerates the degradation of the stationary phase resulting in high column bleed and activity, and finally, short column life.

Lifetime, ease of piercing, sliver formation, bleed and cost can all be important considerations when selecting septa. Most of these characteristics are determined by the material used to made the septa and the method of septa usage.

7.10.2   Septa Hardness

There is a range of septa hardness. The softer septa are easier to pierce with the syringe needle. This makes injecting samples slightly easier and reduces the frequency of bent syringe needles. Softer septa makes it easier to obtain a leak free seal, but it is more difficult to determine when the septum nut is on too tight. Over tightening a septum nut reduces septum life and increases the severity of any leaks.

Softer septa have the tendency to produce tiny slivers of septa upon injection. These tiny pieces of septa fall into the injector. Depending on the design of the liner and injector, these particles can interfere with the flow of carrier gas out of a split line or through a liner. Tailing or broad peaks, irregularities with the split vent flow, or baseline disturbances may occur when septa particulates accumulate in the injector. The worst problem occurs if one of the particles falls into the column. The probability of this problem occurring increases dramatically with larger diameter columns. Severe tailing of nearly all peaks occur if solid particles become lodged in the column.

7.10.3   Septa Bleed

Septa contain a small amount of volatile materials originating from the manufacturing process. These compounds are released from the septum when it is heated to high temperatures in the injector. Many of them accumulate in the column during the times when it is held at cooler temperatures. The longer the column is maintained at the lower temperature, the more septum bleed materials accumulate. When the column is heated such as during a temperature program, the septum bleed materials chromatograph much like an injected sample. The resulting chromatogram can take on a variety of appearances. Usually there will be some type of wide hump or blob often associated with sharper peaks. The size and appearance of the bleed depends on the brand of septum, the injector temperature, the column, the oven temperature, time left at the lower temperature, and the rate and amount of temperature elevation. The amount of septum bleed usually decreases slowly with time - the septum becomes conditioned upon heating. Most of the bleed reduction occurs 15-60 minutes after installation.

The type of injector influences the severity of a septum bleed problem. Injectors that use a high flow rate of carrier gas (e.g., split and splitless) reduces the amount of septum bleed materials reaching the column. Most of the bleed materials are swept out of the injector via the split line during all or most of the analysis. Injectors such as Megabore direct without any split flow direct all of the bleed products into the column. The presence of a septum purge flow helps to reduce septum bleed problems. The proper use of the septum purge function helps to keep the bleed materials from reaching the column. Higher flow rates sweep away more bleed materials, but a loss of the more volatile sample compounds via the septum purge line may occur.

GC systems operated at more sensitive levels respond to the septum bleed materials to a greater extent. Any baseline disturbances appear to be quite severe and can overwhelm small peaks. Some selective detectors do not respond to septum bleed compounds, thus no septum bleed may be seen even though the bleed compounds are present. GC/MS poses an additional problem with septum bleed. The mass spectrum of many of the septum bleed compounds appear to be very similar, if not identical, to many non-polar stationary phase bleed mass spectra. It is easy to misidentify the septum bleed compounds as stationary phase bleed compounds. The presence of peaks, humps or blobs rules out the possibility of column bleed being responsible for the baseline disturbances.

In general, higher quality (i.e., higher cost) septa have substantially lower bleed than low cost ones. At injector temperatures of about 200°C or less, the differences in bleed between the various septa are negligible. As the injector temperature becomes higher, especially above 300°C, the differences in bleed become quite apparent. The higher cost for low bleed septa are well worth the price when operating at high injector temperatures.

Septum bleed materials are only a few of the many substances that can accumulate in a column at low temperatures. The elution of the other materials have the same appearance as septum bleed, thus it is difficult to determine whether the baseline disturbances arise from septum bleed problems. There is a method to determine whether the septum is the source of the contaminants. Before proceeding, make sure the injector and liner are clean. Wrap one layer of aluminum foil around a new septum. Install the septum by placing the side with the continuous piece of foil towards the injector. Be careful as to not expose any of the septum by tearing the foil. Setup the injector and column with the previous conditions. Leave the septum in place for at least 3-4 hours, but do not make any injections. Run a temperature program as previously used and collect the blank chromatogram. Compare with the previous baseline problems. If the severity of the baseline disturbances are noticeably better, the septum is probably the source of the contaminants. If the problem is about the same, the septum is not at fault and the source of the contaminants is elsewhere.

There are several ways to keep septum bleed interferences to a minimum. Lower injector temperatures reduce the amount of septum bleed;

however, the injector still needs to be hot enough to thoroughly and rapidly vaporize the sample. Maintaining the column at higher temperatures reduces the amount of materials that accumulate in the column. This decreases the severity of the baseline disturbances, but at the expense of reduced column life. A finned septum nut keeps the septum at a lower temperature. This reduces the bleed levels, but this type of septum nut is not available for many models of GC's. Pre-treating the septum with solvent or heat helps to remove some of the contaminants, but this can reduce the life of the septum.

7.10.4   Handling Septa

There are several method to clean and condition septa prior to use. Septa can be soaked in solvents to extract some of the contaminants from the surface. This removes contaminants and sometimes reduce septum bleed. Submerging the septa for 15-30 minutes in a beaker of solvent is usually sufficient. Halogenated solvents such as chloroform or methylene chloride should be avoided since many septum swell in these types of solvents. Thoroughly dry the septa before use by heating for 10-15 minutes at 50-100°C. Septa can be preconditioned by baking for several hours at or slightly above the injector temperature. Do not exceed the upper temperature limit for the septa. Placing the septa in a clean beaker and heating them in the GC oven is the easiest method. Sometimes the septa are left in a beaker in the GC oven while it is being used for analyses. A septum is removed as needed for replacement. If septa are heated or solvent extracted before use, reduced septum life is usually experienced.

Septa do not require special handling during installation or storage. The use of clean gloves or tweezers is not required, but will not cause any harm. As long as they are clean, bare hands can be use to handled septa without any problems. Residues from materials such as lotions, food and cosmetics can contaminate a septum so care must be taken to ensure there are none present. The small amount of materials transferred to the septum upon handling quickly vaporize and leave the injector before the next analysis is run. In addition, only a small area of the septum is exposed to the interior of the injector, thus very little of the contaminants will be transferred.

7.11   Injector Maintenance
7.11.1   Cleaning Injectors

Injectors require periodic maintenance and cleaning. They become fouled with non-volatile materials and backflashed materials condensed in cool areas. Seals and gaskets need periodic replacement.

Regardless of the types of samples injected, injectors eventually need cleaning. Samples containing large amounts of non-volatile materials foul injectors much quicker, thus more frequent cleaning is needed. Frequent occurrences of backflash require frequent injector cleaning, also. Baseline disturbances, ghost peaks, tailing peaks or reduced size peaks are the most common symptoms of a dirty injector.

Cleaning an injector involves more than just replacing the liner. Cool the injector to room temperature. Remove the liner, dismantle the various fittings, and remove any of the seals, washers or gaskets. Place a beaker inside the GC oven directly below the injector opening. Rinse the interior of the injector with 20-30 mL of several different solvents. Methanol, acetone and hexane remove nearly all contaminants from the injector; however, other organic solvents will not harm the injector. If biological samples such as blood, urine or plant extracts have been injected, water should be used as the first solvent. After each solvent has passed through the injector, scrub the injector with a small test tube brush. Rinse the injector with an additional 10 mL of the solvent after each scrubbing. Rinse any of the reusable injector parts with the same solvents. Ultrasonication of the parts may help remove contaminants from small crevices and ridges. Thoroughly dry all of the parts before reassembling the injector. New seals or gaskets should be used each time the injector is reassembled. Heat the injector and establish gas flows before installing a column.

In situations where backflash has occurred, the gas lines may need cleaning. Take apart the injector and remove all of the fittings, seals, washers or gaskets. Place a beaker inside the GC oven directly below the injector opening. Label each gas line and disconnect it at the fitting closest to the injector. Some of the lines may be up to 0.5 meters long. Use a large syringe (an adaptor to connect the syringe hub to the line fitting will be necessary) and flush each gas line with 5-10 mL of several different solvents. Methanol, acetone and hexane removes nearly all contaminants from the lines; however, other organic solvents can be used. Do not get any solvents in the flow controllers, pressure regulators or solenoids since this usually damages them. After the last solvent, flush each line with carrier gas for 5-10 minutes. Heat the lines with a hair dryer or heat gun to remove the last traces of solvent. Reattach the lines and reassemble the injector. Heat the injector and establish gas flows before installing a column. In cases of severe contamination, several cleaning may be necessary.

Sometimes it is easier to remove the entire injector from the GC. Consult the GC instruction manual on the proper method to remove the injector. Flushing the lines may be easier with the injector out of the GC. If the injector is unattached, a HPLC or similar type of pump can be used to flush solvents through the lines. This is easier than using a syringe and can be left unattended. To further facilitate cleaning, the injector can be baked out at a high temperature by placing it in an oven. If possible, have a flow of gas through the lines while heating. This improves and speeds up the cleaning process.

To facilitate injector removal, cut the gas lines about 5-10 cm from the injector body. Leave enough tubing to install a stainless union and still have enough slack to move the tubing around. The injector can be easily removed by disconnecting the unions. Forcing the gas lines through the tangle of other tubing, fittings and flow controllers is avoid. Color coding the unions or using an unique union on each line prevents connection of the wrong gas lines.

7.11.2   Cleaning Injector Liners

Depending on the types of samples injected, injector liners need periodic cleaning or replacement. The analysis of dirty samples or active compounds (e.g., -NH and -OH containing compounds), especially at low levels, requires more frequent liner maintenance. It is primarily an economic factor that determines whether liners are cleaned and reused or replaced with a new one. Eventually, all liners have to be replaced or resilylated.

A liner without visible deposits or discoloration is cleaned by rinsing with solvents. A squirt bottle or pipet can be used to rinse the inside of the liner with 10-20 mL of methanol, acetone and hexane. If there are deposits or discolorations, the liner can be ultrasonicated to remove the materials. If desired, a soft bristled brush can be used to scrub the inner surface. It is important not to scratch the surface of the liner. This will expose untreated glass which will adsorb active compounds. If the liner is scratched, it has to be discarded or resilylated. Liners with complex inner surfaces are much harder to clean than simple straight tube types of liners.

A liner becomes more active with use. The surface is longer deactivated and the liner is not suitable for use with most samples. Samples containing active compounds are much more susceptible to liner activity problems. Low concentration samples magnifies the problem. Inactive compounds such as hydrocarbons are fairly immune to liner activity problems. Liners exhibiting unacceptable activity for the analysis of active compounds may have adequate performance for inactive compounds.

When a liner becomes active, it needs to be replaced or resilylated. Resilylation requires an investment in time and reagents, but liners can be recycled. Replacement is easier and does not require handling of solvents and reagents. The cost of new liners, labor expenses, frequency of liner replacement and complexity all have to be considered when determining whether cleaning, resilylation or replacement makes more sense.

7.11.3   Silylating Injector Liners

Untreated glass contains silanol groups that interact with most compounds. These silanols contribution to peak tailing or adsorption problems for active compounds. Silylation converts the silanols to groups that do not interact with active compounds. Before the liner can be silylated it needs to be properly cleaned and treated.

Sample residues, trace metals in the glass matrix and the action of previous silylations need to be removed before a liner can be silylated. This is done with an acid cleaning or leach step. Place the liner in a clean test tube. Cover the liner with a 1N HCl or HNO_3 solution and soak for at least 8 hours (overnight is the most convenient). If the acid solution is highly discolored after use, replace it with clean acid solution and continue to soak until color change is noted. Do not leave the liner is the acid solution for more than 24 hours. Carefully decant the acid solution and thoroughly rinse the liner with deionized water followed by methanol. Dry the liner at 100-150°C. Do not exceed 150°C.

The acid solution removes prior silylation treatments and converts the surface to all silanols groups. These free silanol groups readily react with silylation reagents to form an inert surface suitable for sample injection. The liner is not suitable for use after acid treatment ; it must first be silylated.

Before silylating liners (or any glass), there are several precautions that must be taken. Silylating reagents (silanes) are very reactive towards small molecular weight alcohols (e.g., methanol, ethanol, i-propanol), primary and secondary amines (e.g., dimethyl- and diethylamine) and water. At no point should the silylation reagents come in contact with these types of compounds. The reagents preferentially react with alcohols, amines or water and the liner deactivation will not be successful. The reagents need to be kept dry and exposure to air kept to a minimum. It is best to flush the container with nitrogen and stored it in a freezer. A Teflon lined cap is essential; rubber or paper decomposes in the presence of most silylating reagents. Silylating reagents are very volatile and flammable. They need to be handled in a hood and kept away from sources of heat.

There are two silylation methods - a gas phase and a solution method. The gas phase method provides a slightly better deactivated liner; however, it is a more difficult and labor intensive procedure. The solution method is much easier; however, the liner is not as thoroughly deactivated. Except for the most demanding conditions (very active compound at low levels), the solution method will be satisfactory.

Gas Phase Silylation Procedure:
1. Place the liner in a glass test tube that can be easily flame sealed.
2. Heat the neck of the test tube until a 2-3 mm opening remains.
3. Add 2-3 drops of diphenyltetramethyldisilazane to the test tube.
4. Immediately flush the test tube with a stream of dry nitrogen or argon.
5. Immediately flame seal the test tube (remember the silane is very flammable).
6. Heat the test tube for 3 hours at 300°C.
7. Allow to cool to room temperature.
8. Carefully open the test tube and thoroughly rinse the liner with pentane or hexane.
9. Dry the liner at 75-100°C.

Note: the liner can be stored in the sealed test tube until needed. Several liners can be placed in one test tube; however, rotate the test tube several times during heating to ensure all surfaces are exposed to the silane gas.

Solution Silylation Procedure:
1. Place the liner in a screw cap test tube.
2. Cover the liner with 10% trimethylchlorosilane or dimethylchlorosilane in toluene.
3. Tightly seal the test tube with a Teflon lined cap.
4. Let stand for at least 8 hours.
5. Remove the liner from the solution and thoroughly rinse with toluene then methanol.
6. Dry the liner at 75-100°C.

Note: the liner can remain in the silane solution for longer than 8 hours. Several liners can be placed in one test tube; however, rotate the test tube several times to ensure all of the liner surfaces are exposed to the silane solution.

It is important to rinse all of the excess silylating reagent or solution from the liner before use. Residual reagent may cause an excessively high baseline that often requires hours to days to stabilize. This baseline often mimics column bleed especially if examined by GC/MS. The silylating reagent will not harm the column except for PEG stationary phases. If the injector or column is contaminated with sample residues, some of the residues may react with the silane. This often converts them into very volatile compounds. These newly formed compounds may elute from the column resulting in a high baseline signal, noise or peaks. Rinsing the liner and cleaning the injector and column usually eliminate the problem.

8   DETECTORS

8.1   Introduction

As compounds elute from the column, they interact with the detector. This creates an electrical signal whose size is related to the amount of the corresponding compound.   The detector electronics (electrometer) processes the signal and sends it to a recording device.   The plot of the signal size versus time results in the chromatogram.

There is no single detector that is best for all GC analyses, this being the reason for the variety of choice.   Each GC manufacturer offers the most common ones, but their design and performance may differ.   Since the principle of operation for each detector is the same, there is a general set of rules and characteristics that apply to all.

8.2   Detector Characteristics
8.2.1   Dead Volume

In an ideal situation, the compounds are detected the moment they emerge (elute) from the column.   Due to the design limitations of some detectors, there is space between the end of the column and the point of detection.   This space is called dead volume.   It requires long times for the carrier gas to sweep the compounds from the end of the column into the detector.   This delay in reaching the detector can lead to broad, tailing or reduced size peaks.   These problems occur for several reasons.

The surfaces between the end of the column and the detection zone are usually metal.   Some compounds react with these hot, metal surfaces. Large dead volumes provide a large amount of surface area for the compounds to contact before entering the detector.   Also, the long times spent in this area provides ample opportunity for these compounds to interact with the metal surfaces.   This type of interaction can result in tailing or reduced size peaks.   If the interaction is very short lived, some of the compound molecules are slightly delayed before reaching the detector. The delay of some of the molecules results in a tailing peak. If the interaction is long lived or permanent, some of the solute molecules never reach the detector. The lower number of compound molecules reaching the detector results in a smaller peak.

For compounds that do not interact with the metal surfaces, another problem may occur. It takes a long time for the carrier gas to sweep all of the solute molecules from the dead volume into the detector. From a detector standpoint, peak width is related to the difference in time when the first and last compound molecules enters the detector. Longer sweep times translate into broader peaks. Compounds that elute from the column in tight, compact bands (i.e., narrow peaks) are converted into wide, spread out bands (i.e., broad peaks) by the presence of excessive dead volume. Dead volumes are reduced or eliminated by the proper installation of the column in the detector. The proper use of makeup gas helps to rapidly sweep the compounds through any dead volume and often improves the peak shapes.

8.2.2 Makeup Gas

Most GC detectors require 30-40 mL/min of gas flow to obtain the best sensitivity and peak shapes. The carrier gas flow rates for capillary columns are well below these desired values. A gas is usually supplied to the detector to makeup the difference between the actual carrier gas flow and the desired detector flow. This gas is called the makeup gas. The makeup gas does not flow through the column, but is added at the exit of the column.

Makeup gas helps to rapidly sweep the compounds from the column through the detection zone. This helps to minimizes the effect of any dead volume. No or too little makeup gas can result in tailing or broad peaks since the compounds are not rapidly swept to the detector. Excessively high makeup gas flows usually do not negatively affect peak shape, but may reduce sensitivity. The makeup gas flow becomes more critical as the amount of dead volume increases. Some detectors have more dead volume than others, thus the magnitude of the peak shape problems are different when the makeup gas flow is incorrect.

Some detectors require a specific gas to function. This gas is often used as the makeup gas, thus serving two purposes. Loss of sensitivity and broad peaks may occur when there is not enough makeup gas. The detector does not function properly without enough gas. Excessively high makeup gas flows usually reduces sensitivity, but improves peak shapes. Smaller peaks occur since the compounds are swept through the detector too quickly for the detector to completely respond to their presence.

The best makeup gas depends on the specific detector. Some detectors can properly function with several different makeup gases. One may provide better performance than another, but often the differences are not large. Using the same type of gas for carrier and makeup is convenient since the same gas cylinder and regulator can be used. The best makeup gas flow rate depends on the specific detector and often differs between GC manufacturers. The GC instruction manual provides recommended flows for the makeup gas. A range of flows is often provided. If the best possible sensitivity or peak shapes are desired, experimenting with the makeup gas flows within the specified range is recommended. Using a test sample (the more like the actual samples, the better), comparing peak shape and sensitivity at several different makeup gas flows helps to determine the best makeup gas flow rate. If long term, consistent detector response is important, the makeup gas flow needs to be set at the same value. Different flows alters detector sensitivity and causes inconsistent detector response problems.

Makeup gas flows are measured with a flowmeter at the exit of the detector. Make sure that all of the other detector gases are turned off. Measuring the total flow of two gases simultaneously then subtracting the flow of one from the total flow can lead to errors in the measurements. Since two flow measurements have to be made anyway, it best to measure each gas separately and avoid potential errors. To protect the column, the carrier gas flow is often left on during detector gas flow measurements. This flow needs to be subtracted from the measured makeup gas flow. The carrier gas flow is usually small and does not contribute any significant error to the makeup gas flow value.

8.2.3 Detector Temperature

A detector's temperature can have a pronounced influence on its performance. Some detectors are very sensitive to changes in their temperature while others perform satisfactorily and consistently as long as the temperature is high enough. The detector has to be hot enough to prevent the condensation of any compounds eluting from the column on the surfaces of the detector. Rapid contamination of the detector occurs at low temperatures. Increased noise, loss of sensitivity and shifts in selectivity or linear range are some of the more common symptoms of detector contamination.

The detector does not have to be maintained at a temperature that is higher than the column temperature. While this is desirable, it is not necessary. It is not required to maintain the detector temperature below the column's upper temperature limit since no significant damage to the column occurs. A darkening of the polyimide coat of the column may be observed at detector temperatures above 325°C. This section of column inside the detector may be slightly weakened if detector temperatures above 350°C are used. Occasional trimming of the section is recommended to prevent breakage of the column in this area during use.

Temperature sensitivity detectors should be kept away from heating and cooling vents or other sources of temperature change. When used at their more sensitive settings, these detector may respond to the sudden change in the ambient temperature. Usually, a sudden shift in the baseline, drift of the baseline or change in the sensitivity is noticed if a sudden temperature change occurs.

8.2.4 Sensitivity

The most basic definition of sensitivity is the smallest amount of a compound that can be detected. The actual amount will vary depending on the compound, detector, peak width, and gas flow. A value called the signal to noise ratio (S/N) is often used to definite sensitivity. The signal is the height of the peak and the noise is the height of the baseline noise (Figure 8-1). There are many different definitions of signal to noise; however, this one is the most practical for chromatographers. Be aware of any scale changes when measuring peak and noise heights. Sometimes, to measure the noise, the vertical scale has to be expanded. The peak is then too large to be measured and the scale has to be reduced to view the entire peak. The scale differences (which may be not linear) have to be accounted for in order to obtain an accurate measurement.

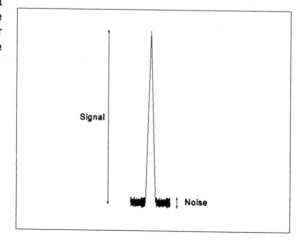

Figure 8-1:   Signal to Noise

The S/N ratio is often used as a measure of a detector's response. For example, 10 pg of a specific compound gives a S/N of 10. Often minimum detection limits (MDL) are quoted as S/N ratios. For example, a peak has to have a S/N greater than 4; a peak with a lower S/N value is too small to be assigned to a compound. Many regulatory agencies set MDL's at a S/N of 2-4. This range is probably more applicable to analytical standards in clean solvents. For the average gas chromatographic detector and normal operating conditions, a S/N value of 4-8 is more realistic for extracted samples. Peak assignments become suspect at very low S/N values. As the complexity (i.e., dirtiness) of a sample increases, the S/N values should be higher to compensate for possible interferences (co-eluting compounds) and higher background.

The response of some detectors change with even small flow rate changes of the carrier or makeup gases. The response of other detectors are fairly independent of the gas flows. Detectors that respond independently of the carrier or makeup gas flow rates are called mass flow rate detectors. The detector response is proportional to the mass of compound passing through the detector per unit time. In theory, these detectors are not affected by small changes in the carrier and makeup gas flow rates. In practice, changes in gas flows greater than 25% have a significant effect on detector behavior. Detectors that are affected by the volume of gas flowing through the detector are called concentration dependent detectors. The detector response is proportional to the mass of solute per unit volume of gas. In other words, the concentration of the compound in the gas. The same mass of compound eluting from the column generates a peak of different size depending on the flow of the carrier and makeup gases. At higher flow rates, the compound becomes diluted by the carrier or makeup gas. This reduces the concentration of the compound, and the peak size thus decreases. Lower flow rates increase the peak size, thus sensitivity improves. Excessively low flow rates create dead volume problems; therefore, there are lower flow limits for each detector. Drifting baselines related to the temperature program are common for concentration dependent detectors. As the column temperature increases during a temperature program, the carrier gas flow rate drops. Since the volume of carrier gas is decreasing, the background signal also changes. Because the baseline is a function of the background signal, this gives rise to a drifting baseline that follows the changes in the temperature program. Mass flow rate detectors do not exhibit the baseline drift behavior since they are not very sensitive to changes in the gas flows.

8.2.5 Selectivity

Only a few detectors respond to every compound eluting from the column. In most cases, only particular types of compounds elicit a response from a detector. For some detectors, this is a very broad range of compounds, and for others, only a narrow range of compounds. A detector is highly selective if it responds to only a narrow range of compounds. This means that peaks are obtained only for compounds whose structures contain the characteristic necessary for interaction with the detector. The other compounds pass through the detector; however, no peaks are obtained for these compounds. A less complex (i.e., fewer peaks) chromatogram is achieved. It is now much easier to resolve the peaks of interest from the matrix or background peaks since there are fewer interfering peaks present.

It is usually an atom, functional group or structural feature that elicits a selective detector's response. For example, a detector responds only to compounds containing halogens; all other compounds have little to no response. Detector selectivity can be viewed as either enhanced response for a specific class of compounds or a lack of response for another class of compounds. Most selective detectors have an enhanced response for one class of compounds and a poor response towards others. Most detectors have substantially better sensitivity for their selective compounds than the general purpose (i.e., not very selective) detectors. A few detectors only filter out the nonselective compounds. Their sensitivity is not much better than some of the general purpose detectors, but the filtering out of the unwanted peaks makes it easier to discern and measure the peaks of interest.

Detector response or behavior may be affected by the presence of nonresponsive compounds. Alterations in the peak size or linear behavior (Section 8.2.6) are the most common problems related to the presence of nonresponsive compounds. Problems usually occur when a high concentration of a nonresponsive compound elutes very near or co-elutes with a responding compound. Even though the detector does not produce a peak response, the detector is influenced in a manner that affects its behavior. In most cases, the peak of interest becomes reduced in size. This effect is often called quenching.

Selectivity is usually measured by comparing the response of the selective compound to a reference compound (Equation 8-1). Hydrocarbons are the usual reference compounds for the highly selective detectors.

8.2.6 Linear Range

Any change in the amount of a compound in an injected sample should result in a corresponding change in the compound's peak size. For example,

$$Selectivity = \frac{Peak\ Size\ of\ Selective\ Compound}{Peak\ Size\ of\ Reference\ Compound}$$

Equation 8-1:   Selectivity

if the amount of compound is doubled, the size of the resulting peak should also double. This behavior occurs only over a certain range of compound amounts - this range is called the linear range of the detector. If the amount of compound is outside this range, the response of the detector does not directly reflect the amount of compound actually passing through the detector.

The linear range of a detector is determined with a simple experiment. A series of standards of the compounds of interest are made over a wide concentration range. The lowest should be at the minimum detection limit (Section 8.2.4), and the highest at least 10^5 times higher. The minimum number of standards should be five with a greater number providing more accurate information. Inject each standard at least 3 times and plot the average of the peak size versus the amount of the compound. The linear range is the region where a straight line can be drawn through the points (Figure 8-2a). Another method is to divide the peak areas by the corresponding compound amounts. These values are plotted and the linear range is the region of the straight line (Figure 8-2b). Usually a deviation of ±5% is allowed before the linear range is considered to be exceeded.

Depending on the column and detector, the column's capacity may be exceeded before the maximum value of the detector's linear range is reached. For some detectors, the linear range varies by compound and operating conditions, thus the linear range for each compound and operating condition may need to be individually determined. All injector conditions need to be kept constant to ensure a true measure of a detector's linear range. Variations can affect the amount of compound entering the column, and reaching the detector. Factors such as split ratios need to be considered also.

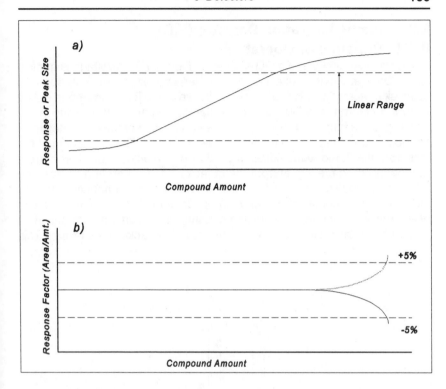

Figure 8-2: Linear Range Curves

When performing quantitative analyses, calibration curves are normally generated from the injection of standards. If the compound amounts in the standards are within the detector's linear range, a three point calibration curve can be used. The straight line drawn through the three points accurately reflects the peak size to compound amount relationship. If the compound amounts are near either end of the linear range, at least a five point curve should be used. The nonlinear response around the ends of the linear range require more points to accurately measure the response relationship.

8.3   Flame Ionization Detector (FID)
8.3.1 Principle of Operation
A simplified drawing of a FID is shown in Figure 8-3. A voltage is applied to the flame jet and the collector. A hydrogen and oxygen (usually supplied as an air mixture) flame is maintained at the tip of the jet. Carrier gas exiting from the column is delivered into the flame. Organic compounds in the carrier gas are burned in the flame. This combustion process creates ionic species which are attracted to the charged collector. The movement of the ions from the flame to the collector produces a small current. This current is measured as the signal from the detector. A background or baseline signal is always present and it is a result of trace levels of impurities such as gas and system contaminants and normal column bleed. The introduction of an organic compound into the flame results in the compound's ionization and an increase in the detector current above the background level.   This

Figure 8-3:   Major Components of a FID

detector signal is processed by the recorder into a corresponding peak in the chromatogram. Greater amounts of the compound generate a larger signal, thus giving a larger peak.

8.3.2 Gases

Hydrogen and air are used as the combustion gases. Usually, the air flow is about 10 times the flow of hydrogen (30-40 mL/min hydrogen and 300-400 mL/min for air). Each model of detector is unique, thus the exact flow values are found in the instruction manual. The flows of the hydrogen and air combustion gases with FID's affects sensitivity. The same flow values are needed to maintain consistent performance. If the gas flows are too far away from the recommended values, difficulties in maintaining or lighting the flame occur.

Nitrogen is a better makeup gas than helium. Using nitrogen as the makeup gas increases detector sensitivity by 10-20% over helium. Recommended makeup gas flow rates are found in the GC's instruction manual. Makeup gas flow rate affects sensitivity, thus a correct and consistent flow is important for best performance. Experimenting with adjustments (5-10%) to the makeup gas flow may improve sensitivity, but often the gains are quite small and not worth the effort. Hydrogen is not recommended as a makeup gas. Since hydrogen is one of the combustion gases, its presence alters the flame behavior. Sensitivity problems and flame lighting difficulties are the most prevalent problems.

8.3.3 Column Position

The proper column position in the detector is essential for optimal performance. The column end needs to be ≤1 mm below the tip of the flame jet. If the end of the column is too far below jet tip, dead volume problems occur. If the column is positioned too far into the jet, difficulty in lighting the flame, peak shape problems or excessive background noise may occur. The GC instruction manual is a good source of information on the positioning the column in the detector.

8.3.4 Temperature

The temperature of a FID is not critical to its performance. The detector needs to be hot enough to prevent the condensation of high boiling compounds. A temperature of 225-325°C is typical. The higher temperature range is recommended to keep the detector from rapidly becoming dirty.

8.3.5 Selectivity

The FID responds to compounds containing a carbon-hydrogen bond. The exceptions are formaldehyde and formic acid which have very poor responses. Fully halogenated hydrocarbons give weak responses even though there are no carbon-hydrogen bonds present. The FID does not respond to permanent gases (O_2, H_2, N_2, He, Ar, Xe, etc.), oxides of nitrogen (NO, NO_2, N_2O, N_2O_2), sulfur compounds (H_2S, SO_2, CS_2, COS, etc.), CO_2, CO, and water. Large solvent peaks are nearly eliminated for samples in CS_2 or water. Baseline disruptions in the area where these compounds elute are common. In practice, CS_2 gives a larger response than expected and can be of appreciable size at sensitive settings.

8.3.6 Sensitivity and Linear Range

The practical minimum detectable quantity is 0.05-0.5 ng of compound. There is an increase in response with respect to the number of carbons - the more carbons atoms, the better the response. Response decreases as the amount of substitution with halogens, sulfur, oxygen, nitrogen and phosphorus atoms increases. For example, chloropropane has a poorer response than propane. The response loss due to substitution is variable and is dependent on the exact groups and their position on the molecule. Carbonyl carbons do not contribute to the response.

A FID is a mass flow rate detector. Changes in the carrier and makeup gas flow rates do not significantly affect the sensitivity. Sensitivity changes do occur; however, large changes in the gas flow rates are necessary.

The linear range of 10^5-10^6 for the FID is among the largest of all GC detectors. This makes the FID suitable for use with a very wide range of sample concentrations. Three point calibration curves are acceptable for FID due to their large linear range.

8.3.7 Verifying Flame Ignition

The signal output as displayed on the data signal or GC display can be used to verify flame ignition. Check the signal prior to lighting the FID. The signal is higher when the flame is burning. By comparing the two values, it is easy to determine if the flame is lit. The actual numerical values depend on the source of the detector signal. Different GC's and data systems report detector output in different units or values. This is not a problem since only a comparison between the lit and unlit detectors are needed.

To physically verify that the flame is lit, place a cool, shiny object (e.g., wrench, knife blade, etc.) over the vent of the detector. After several seconds, condensation should be visible on the shiny object after it is removed from the vent. If no condensation is visible, the detector flame may not be lit.

If there are doubts about whether the flame is lit, inject about 1 µL of a detectable solvent. A large peak should be obtained after a few minutes providing the column temperature is high enough to elute the solvent. The solvent will not harm the column even if it has not been conditioned.

8.3.8 Maintenance

FID's require little maintenance to keep them performing at satisfactory levels. The hydrogen, air and makeup gas flows should be occasionally measured. They can drift over time or be unintentionally changed without knowledge of it occurring. Each gas flow should be measured independently to obtain the most accurate values.

The FID requires periodic cleaning. In most cases, this only involves the collector and the jet. FID cleaning kits are available containing brushes and wires that simplify the cleaning of all of the detector parts. The brushes are used to dislodge particulates clinging the metal surfaces. A fine wire is used to clean the jet opening of particles. Do not force too large a wire or probe into the jet opening or the opening will become distorted. A loss of sensitivity, poor peak shape and/or lighting difficulties may result if the opening is deformed. The various parts can be ultrasonicated after cleaning with a brush. Eventually the jet needs replacing, but this should be a fairly infrequent necessity.

8.3.9 Common Problems
8.3.9.1   Change in Sensitivity

A loss of sensitivity is the result of a loss in peak size and/or an increase in the amount of noise. Losses of peak size or increases in noise are usually from improper detector gas flows or a dirty detector. Sensitivity is dependent on the flow rates of hydrogen, air and the makeup gas. Incorrect flow rates alter FID sensitivity. In some cases, only the peak size or the noise is affected. Some compounds may suffer a larger response shift than others.

Combustion and makeup gases that are improperly set can cause a loss of peak size. An increase in the noise may occur when the combustion gas flows are incorrect, but this usually occurs only when the gas flows are significantly away from the proper flow rates. A dirty detector may affect sensitivity also. Deposits form on the jet and collector which may partially inhibit the formation or detection of the ions. A reduction in the number of ions generated or detected results in a loss of response. Contaminants in the detector can contribute to the background noise. They slowly reach the collector and elevate the background signal.

For collectors that can be disassembled, an inadequately tightened collector may cause noise. Some jets have an opening that is larger than the outer diameter of the column. If the column is pushed too far into the detector, the column can protrude into the flame. The polyimide coating decomposes and causes high background signals and noise problems. This is evident from the melted end of the column, the lack of polyimide coating (clear glass is visible) or black discoloration on the end of the column upon removal from the detector. Usually peak shape problems occur also.

Contaminated detector gases or gas lines can cause excessive noise problems. These problems are hard to distinguish from other sources of contamination, thus carefully examination of the GC and related devices is required. The use of a jumper tube test may help to isolate the source of the contamination.

8.3.9.2   Difficulty in Lighting the Flame

Difficulties in lighting the FID flame can usually be attributed to improper gas flows or a dirty jet. The incorrect mixture of hydrogen and air will not sustain a flame. A popping noise may be heard, but this does not mean the flame is lit. The proper ratio is needed to maintain a stable flame. Sometimes, if the makeup gas flow is severely off, the flame will not light. A dirty or clogged jet may prevent the proper flow of hydrogen from reaching the combustion area. Particulate matter or a column completely blocking the jet opening may prevent hydrogen from flowing at the proper rate.

Make sure that the FID ignitor is working. It usually is located and visible near the top of the FID. It should have an intense orange glow to it when on. Lighting the flame with a match is not recommended since foreign matter may fall into the detector giving rise to noise. A lighter is a better choice of alternative ignition.

Hydrogen carrier gas is used at high flow rates. For Megabore columns, the flow of hydrogen can be quite high. This hydrogen is additive with the detector hydrogen and alters the total amount of hydrogen reaching the combustion area. This may be enough of a change to make it difficult to light the flame. Reducing the flow of detector hydrogen by the same amount as the column flow usually solves this problem.

Some FID's become difficult to light with use. Sometimes it is a specific jet that causes the problem. If no problems are evident, it may be just a quirk with that particular FID. A difficult to light FID can sometimes be coaxed into lighting. Lower the air flow to around zero. Activate the ignitor then slowly raise the flow of air. Usually a popping sound will be heard, but continue to raise the air flow to its recommended value.

8.3.9.3   Peak Shape Problem Attributed to the Detector

Improper column installation in the detector or too low makeup gas flows are the most common source of broad or tailing peaks caused by the detector. If a column is forced too far into the detector, a broken column end may result. Upon tightening the column nut, the column is forced upward and its end may be crushed in the tip of the jet. The broken column end and resulting debris disrupts the flow at the end of the column. This turbulent flow may cause broad or tailing peaks. When a column is too far away from the tip of the jet, there is excessive dead volume. Tailing or broad peaks may result with the problem being more severe for very volatile compounds. Too low or nonexistent makeup gas flows often causes tailing or broad peaks.

It is possible to push the column through the opening of some FID jets. In addition to excessive noise, peak shape problems may also occur. Depending on the design of the FID and jet, it may be possible to miss the jet. The column slides around the outside of the jet instead of into it. This creates a tremendous amount of dead volume and the usual occurrence of tailing peaks.

8.3.9.4   Miscellaneous

A drifting baseline is sometimes obtained when using hydrogen as the carrier gas. This occurs with temperature programs and is more common at very high flow rates. The carrier gas flow decreases as the column temperature increases. This changes the amount of total hydrogen (column and combustion) in the detector. This change causes a slight shift in detector baseline or background signal. The baseline shift is more severe when there is a large difference between the initial and final temperatures in a temperature programmed run. Also, the shift becomes more visible when using more sensitive settings.

Spikes in the chromatogram can be caused by particulate matter falling into the detector. This generates a sudden surge in the signal, and thus a spike in the chromatogram. Particulates usually accumulate on the collector or upper regions of the FID.

Large volumes of nonflammable solvents may extinguish the FID flame. This is a particular problem with water as the sample solvent. If this problem occurs, changing the sample solvent is the easiest solution. In cases where a change is not possible, the options are limited. Sometimes changing the hydrogen and air flows may help to keep the flame lit. Usually increasing the amount of hydrogen is the most successful. Adding a small amount (5-10%) of methanol or acetone to the sample may help. These are flammable solvents and may help to sustain flame combustion in the presence of water. Sample concentrations have to be adjusted to account for the increase volume of the sample. Also, a larger solvent front is obtained due to the much better response of the added solvent.

8.4   Nitrogen-Phosphorus Detector (NPD)
8.4.1 Principle of Operation

A simplified drawing of a NPD is shown in Figure 8-4. A voltage is applied to the jet and the collector. A bead or cylinder (usually rubidium silicate) resides above the jet. The bead is electronically heated to 600-800°C by application of a current to the bead. A plasma is sustained in the region of the bead by the addition of hydrogen and air. The flow of hydrogen is too low to support a flame. Nitrogen or phosphorus containing compounds react in the gaseous boundary layer around the surface of the bead to produce specific ions. A small current is produced by the movement of these ions from the plasma to the charged collector. This current is measured as the signal from the detector. A background or baseline signal is always present, and it is a result of trace levels of impurities such as gas and system contaminants and normal column bleed. The introduction of a nitrogen or phosphorus containing compound results in its ionization and an increase in the detector current above the background level. This signal is processed by the recorder into a corresponding peak in the chromatogram. Greater amounts of the compound generate a larger signal, thus giving a larger peak.

Since a flame is not sustained in a NPD, hydrocarbon ionization does not occur as with a FID. This results in the high selectivity to nitrogen and phosphorus containing compounds and the lack of response for other compounds. The ionization efficiency of a NPD is about 10,000 times better than a FID. This means that the sensitivity of a NPD for nitrogen and phosphorus compounds is about 10,000 times better than a FID.

Figure 8-4: Major Components of a NPD

Some older NPD's could not efficiently detect nitrogen and phosphorus compounds at the same time. For example, in the phosphorus mode, the response for nitrogen compounds was about 10 times less than in the nitrogen mode. Most NPD's made since the early 1980's can simultaneously operate in both modes.

8.4.2 Gases

Hydrogen and air are used as the gases to support the plasma. Flow rates of 2-6 mL/min for hydrogen and 60-200 mL/min for air are typical. NPD selectivity and sensitivity is very dependent on the hydrogen flow rate; the flow of air is less critical. Older NPD's use a different hydrogen flow for the nitrogen and phosphorus modes (1-5 mL/min for nitrogen and 20-30 mL/min for phosphorus) and 70-100 mL/min of air.

Helium is the usual makeup gas for NPD's since it provides the best sensitivity. The flow rate is usually 30-40 mL/min. The GC's instruction manual recommends the best flow rates for a specific model of NPD. Sensitivity, especially for nitrogen compounds, is affected by the makeup gas flow rate. Too low or no makeup gas results in very poor peak shapes and sensitivities. Nitrogen can also be used as a makeup gas. Actually, it provides more stability to the NPD.

8.4.3 Column Position

The proper column position in the detector is essential for optimal performance. The column end needs to be ≤1 mm below the tip of the jet. If the end of the column is too far below jet tip, dead volume problems occur. If the column is positioned too far into the jet, peak shape problems or excessive background noise may occur. The GC instruction manual is a good source of information on the positioning the column in the detector.

8.4.4 Temperature

Selectivity and especially sensitivity are affected by the detector temperature. A stable temperature environment is critical to maintain consistent NPD behavior. The detector temperature for a NPD should be as high as reasonably possible. Temperatures of 275-325°C are routinely used. High temperatures help to stabilize the NPD. Large concentrations of a compound, such as a sample solvent, often create large baseline disturbances. A high detector temperature helps to minimize the severity and length of the baseline disturbance. A high detector temperature is also recommended when high carrier or makeup gas flows are used. The higher flow of gas has a tendency to cool the interior of the detector. This may reduce detector sensitivity; a higher temperature can thus be used to compensate for the loss of response. Higher detector temperatures allow the use of lower bead currents, thus increasing the life of the bead.

8.4.5 Selectivity

The selectivity of a NPD for nitrogen over carbon is 10^3-10^5 and for phosphorus over carbon is 10^4-$10^{5.5}$. Phosphorus specificity is easily obtained over a relatively wide range of gas flow rates and bead currents. Higher hydrogen flows do increase the specificity of phosphorus over nitrogen. Nitrogen specificity is significantly more dependent on gas flow rates and bead current. For this reason, the NPD is usually optimized for nitrogen specificity.

8.4.6 Sensitivity and Linear Range

The practical minimum detectable quantity is 0.1-0.5 pg for phosphorus compounds and 0.5-1 pg for nitrogen compounds. The sensitivity is primarily dependent on compound structure, bead current and age, gas flow rates, and detector temperature. There are known instances where sensitivity is very dependent on compound structure. The response for compounds with the nitrogen atom bound to a carbon with a hydrogen is better than those without (e.g., amides, nitro compounds). Operating the bead at high currents produces the best sensitivity, but at the expense of increased noise and reduced bead life. As the beads become older, a higher current is usually necessary to maintain previous sensitivity levels. Eventually, the noise become too high and renders the detector unusable. Consistent gas flows, especially hydrogen, are necessary for constant response. Small flow changes can result in large sensitivity changes primarily for nitrogen compounds. Increasing the hydrogen flow into the higher recommended flow range tends to increase the phosphorus compound sensitivity, but usually at the expense of decreased bead life. A higher detector temperature usually provides slightly better sensitivity especially at high carrier and makeup gas flow rates.

A NPD is a mass flow rate detector. Small changes in the carrier and makeup gas flow rates do not significantly affect the sensitivity. Nitrogen compounds are much more susceptible to flow related response changes than phosphorus compounds. Larger changes in the gas flows affect the temperature in the detector. It is this temperature change than mainly affects sensitivity.

The linear range for a NPD is typically 10^4-10^5. This range is dependent on the compound, gas flows, bead current and age, and detector temperature. Five point calibration curves are recommended with NPD's (due to their temperamental nature) even though their linear range is good.

8.4.7 Maintenance

NPD's are temperamental and require frequent maintenance. Small changes in any of a number of parameters can significantly change the performance characteristics of a NPD. The bead requires the most maintenance. It needs changing frequently, thus a spare is a necessity. The beads have to be kept dry which limits their storage life to about 6 months.

When a new bead is installed, slowly raise the detector temperature and bead current. Rapid heating can crack or break the bead especially if it has been stored under humid conditions. It has been observed that higher hydrogen flows and bead currents decrease bead life. If the NPD is not in use, the hydrogen flow and bead current should be reduced or turned off to increase bead life. Make sure there is some type of gas flow in a heated detector or when there is current to the bead.

The hydrogen, air and makeup gas flows should be frequently measured. They can drift over time or be unintentionally changed without knowledge of it occurring. Each gas flow should be measured independently to obtain the most accurate values. NPD's are very sensitive to changes in the gas flows and consistent flows are necessary to maintain performance levels.

The NPD requires periodic cleaning. In most cases, this only involves the collector and the jet. Cleaning kits are available containing brushes and wires that simplify the cleaning of all of the detector parts. The brushes are used to dislodge particulates clinging the metal surfaces. A fine wire is used to clean the jet opening of particles. Do not force too large a wire or probe into the jet opening or the opening will become distorted. A loss of sensitivity or poor peak shape may result if the opening is deformed. The various parts can be ultrasonicated after cleaning with a brush. Eventually the jet will need replacing, but this should be a fairly infrequent necessity.

8.4.8 Common Problems
8.4.8.1   Change in Sensitivity

NPD's are noted for their instability with the sensitivity being the most variable feature. Practically every detector parameter affects the sensitivity. Due to the erratic nature of the bead, the sensitivity can frequently and rapidly change without warning or alteration of any of the operating parameters. Most cases of sensitivity changes are due to some type of change in the bead.

When the detector is first turned on, the sensitivity and the signal level changes slowly over several hours. For applications that require very stable operation (e.g. trace level analysis), leave the detector on overnight for maximum stabilization. If the signal continues to drift, the bead may need replacing.

Sensitivity shifts occur with flow rate changes in the detector gases especially the hydrogen. Consistent flows are necessary to maintain constant or stable sensitivity. Depending on the compound and the direction of the flow rate change, the sensitivity may increase or decrease. Large changes in the gas flows change the temperature of the detector. This shift in detector temperature causes a change in the sensitivity. Obviously, a change in sensitivity is seen when the detector body is heated to a different temperature.

The primary cause of sensitivity changes is the bead. The gradual but normal depletion of the bead results in a drop in the sensitivity. Usually the bead current is increased to compensate for the loss. Increasing the current decreases bead life. Eventually the bead is depleted to a point where the sensitivity is inadequate or the noise is too high.

Another bead related sensitivity problem is due to contamination of the bead. Detector response is strongly influenced by the amount of surface contamination of the bead. Common sources of detector contamination are high boiling sample residues, fingerprints and leak detection fluids. Excessive column bleed products can coat the bead with contaminants and reduce the sensitivity. Excess silylation (derivatization) reagents in samples should be avoided for the same reason. Scratched or damaged beads adversely affect sensitivity and increase noise levels.

Avoid exposing NPD samples to glassware that have been washed in phosphorus based detergents. The residue phosphorus compounds can be detected by the highly sensitivity NPD. High baseline signals or noise can occur with this background of phosphorus compounds.

8.4.8.2   Peak Shape Problem Attributed to the Detector

Improper column installation in the detector can cause broad or tailing peaks. If a column is forced too far into the detector, a broken column end may result. Upon tightening the column nut, the column is forced upward and its end may be crushed in the tip of the jet. The broken column end and resulting debris disrupts the flow at the end of the column. This turbulent flow may cause broad or tailing peaks. When a column is too far away from the tip of the jet, there is excessive dead volume. Tailing or broad peaks may result with the problem being more severe for very volatile compounds.

Too low or nonexistent makeup gas flows may cause tailing or broad peaks. The error in the makeup gas flow also affects sensitivity, thus a loss of peak size accompanies a peak shape problem due to makeup gas flow problems.

Peak tailing can sometimes be caused by the bead, especially if dirty samples are frequently injected. Older beads are also more susceptible to this problem. Phosphorus compounds appear to be more sensitive to bead induced peak tailing. Increasing the bead current or the hydrogen flow often reduces the amount of tailing; however, a shift in sensitivity usually occurs.

8.4.8.3 Baseline Problems

A drifting baseline is obtained when using a temperature program. The problem is more severe with hydrogen as the carrier gas. The carrier gas flow decreases as the column temperature increases. This changes the amount of total gas in the detector. If helium is the carrier gas, a temperature change occurs in the detector. If hydrogen is the carrier gas, a change in the hydrogen flow and the temperature of the detector changes. This change causes a shift in detector baseline or background signal. The baseline shift is more severe when there is a large difference between the initial and final temperatures in a temperature programmed run. Also, the shift becomes more visible when using more sensitive settings. There is very little that can be done to compensate for this problem. The only solution is to use a pressure programmable injector in the constant flow mode (Section 7.6.1) to maintain a constant flow through the temperature program.

Some stationary phases contain nitrogen. The degradation products from these stationary phases usually contain nitrogen containing species. Normal column bleed (Section 4.6) is amplified due to the extreme sensitivity of the NPD to nitrogen. The baseline rise appears to be severe and excessive when it is normal. As a general rule, nitrogen containing stationary phases should not be used with NPD's.

Large concentrations of a compound can cause a baseline disturbance. This is most common around the solvent front. Negative baseline drops and erratic baselines are common immediately after the solvent front. This even occurs with solvents that are not detected by the NPD. Temporary changes in the detector temperature or bead behavior are the cause of the baseline disturbance.

If the bead is severely contaminated, it is possible to get negative peaks for hydrocarbons. If a negative peak occurs in the same location, especially for multiple injections of the same sample, it may be related to bead contamination. If the negative peak occurs near the solvent front, it may be only the effect of the high concentration of sample solvent.

8.5 Electron Capture Detector (ECD)

8.5.1 Principle of Operation

A simplified drawing of an ECD is shown in Figure 8-5. A source of electrons from a radioactive source bombards the column effluent passing through the detector. The most common source is ^{63}Ni supported by a metallic foil. An auxiliary or moderating gas is added to the detector and in many cases functions as the makeup gas as well. Nitrogen or a 95:5 ratio of argon/methane (sometimes referred to as P5) is used as the auxiliary gas. Both gases can be ionized by the free electrons from the source. The electron bombardment of the auxiliary gas creates a plasma containing, among other species, thermal electrons. A potential difference is applied to the detector cell which allows the capture of the negatively charged, thermal electrons by the anode. The migration and capture of the thermal electrons creates a background current which is the baseline (background) signal. When a molecule with the ability to capture an electron elutes from the column and enters the detector cell, it captures a thermal electron. The loss of thermal electrons decreases the current in the cell. The amount of the current loss is measured and becomes the detector signal. The signal is processed by the recorder into a corresponding peak in the chromatogram. Greater amounts of the compound generates a larger signal, thus giving a larger peak.

Figure 8-5:   Major Components of a ECD

8.5.2 Gases

Nitrogen or argon/methane (95:5) is necessary for an ECD. This gas is called the auxiliary or moderating gas and is often supplied as the makeup gas, thus serving two purposes. In general, most ECD's require a total gas flow rates of 20-60 mL/min. If they are supplied separately, the flows of the auxiliary and makeup gases need to be considered as a total flow. A change in either gas flow significantly affects the ECD's behavior. The exact gas flow rates are found in the GC's instruction manual. ECD's are very sensitive to changes in the flow rate of auxiliary or makeup gas. Changes as little as 1-2 mL/min can be very noticeable. Constant flows are necessary to maintain consistent performance from an ECD.

Better sensitivity is obtained if nitrogen is used as the auxiliary gas. The sensitivity is approximately two times greater than possible with argon/methane. Argon/methane provides slightly better linear range at the sacrifice of sensitivity. The choice of the auxiliary moderating gas primarily depends upon the requirements of the analyses being performed. Samples with possible wide concentration ranges may benefit from the better linear range available with argon/methane. High sensitivity requirements makes nitrogen a better choice. Another consideration is that nitrogen is less expensive per cylinder than argon/ methane.

To achieve the best overall results for sensitivity and peak shape, a small amount of experimentation is recommended. Adjust the auxiliary gas flow rate by about 10% above and below the recommended flow values in 2-3% increments. A change in the sensitivity is usually observed. The peak widths narrow as the flow rates increase. Choose the value where the best sensitivity is obtained without a tremendous sacrifice of peak width. For some ECD's, a change of only 1-2 mL/min in the auxiliary gas flow can significantly affect the sensitivity or peak width without any noticeable changes in the other.

Helium is the preferred carrier gas for ECD's. Hydrogen is not recommended by some GC manufacturers since it may reduce the usable lifetime of the detector. If nitrogen is used as the carrier gas, sensitivity changes and baseline drifts become quite evident when using a temperature program. The flow rate of nitrogen decreases as the column temperature increases. The amount of nitrogen reaching the detector changes during the run, thus the change in the sensitivity and baseline.

8.5.3 Column Position

The position of the column relative to the detector cell is important for optimal sensitivity and peak shape. The end of the column should be at the entrance to the cell. If the column is positioned too low, dead volume problems occurs. It is easy to insert the column into the cell for many ECD's. Insertion of the column into the cell itself normally results in the loss of sensitivity, and in extreme cases, some distortion of peak shape. The instrument manual is the best source of information on positioning the column in the detector.

8.5.4 Temperature

Most ^{63}Ni detectors have a maximum temperature limit of 350-400°C. They are usually operated in the higher temperature range. ECD temperature affects sensitivity. Sensitivity usually increases with an increase in the ECD temperature. There are exceptions with carbonyl compounds being the most common examples. Using higher ECD temperatures has the added benefit of reducing detector contamination caused by the condensation of high boiling sample components.

8.5.5 Selectivity

ECD's are selective for halogenated compounds, nitrates, conjugated carbonyls and certain organometallic compounds. Poor response is obtained for hydrocarbons, alcohols, amines, aliphatic substituted aldehydes, thioethers and aromatics. The best selectivity is for halogenated compounds. Table 8-1 lists some relative responses for a variety of compounds. ECD's are primarily used for the analysis of halogenated compounds due to their large selectivity over most other compounds.

8.5.6 Sensitivity and Linear Range

The practical minimum detectable quantity is 0.1-10 pg for multiply substituted halogenated compounds, 10-100 pg for nitrates and mono halogenated compounds and 50-250 pg for carbonyl compounds. Halogen response increases in the order of I > Br > Cl > F. ECD response increases with multiple halogen substitution especially on the same carbon atom. The sensitivity trends concerning the type, number and position of the halogens are readily evident in Table 8-1.

The ECD is a concentration dependent detector meaning sensitivity decreases at higher flow rates of carrier, makeup and auxiliary gases. It appears that low gas flows are preferred to maximize sensitivity. To obtain desired sensitivities, ECD's require a large cell volume, and thus a high total

Table 8-1:   Relative Response on ECD

Compound	Relative Response Compared to Benzene
Benzene	1
Acetone	8
Di-n-butylether	10
Methyl butyrate	15
1-Butanol	17
1-Chlorobutane	17
1,4-Dichlorobutane	250
1,1-Dichlorobutane	1250
1-Bromobutane	5×10^3
1-Iodobutane	2×10^6
Chlorobenzene	1250
Bromobenzene	7500
Chloroform	1×10^6
Carbon tetrachloride	7×10^6
Chlorotrifluoromethane	100
Dichlorotrodifluoromethane	9×10^5
Trichlorofluoromethane	4×10^7
Bromotrifluoromethane	3×10^6

gas flow to effectively sweep the cell. Optimal sensitivity and minimal peak broadening can not be achieved at the same flow rates; therefore, a compromise must be reached. High flow rates that sacrifice some sensitivity are used to maintain narrow peak widths.

The linear range of ^{63}Ni ECD's is poor and barely exceeds 10^3 for some ECD's and compounds. Argon/methane as the auxiliary gas results in a slightly better linear range than with nitrogen. This linear range gain is made at the sacrifice of sensitivity. Nitrogen provides about two times the sensitivity obtained with argon/methane. It is very easy to exceed the linear range for an ECD. Five point calibration curves are recommended since samples are outside the linear region of the ECD's response curve are fairly common.

8.5.7 Maintenance

ECD's require little maintenance. Due to the slightly radioactive source, the cell is sealed and is not directly accessible. ECD's have a lifetime of 2-5 years. After that period, the detector has to be returned to the GC manufacturer for rebuilding or replacement. Baseline or sensitivity problems are the most common symptoms of an old ECD. The types of samples injected and the amount of oxygen exposure affects ECD life. Lower oxygen exposure and cleaner samples will prolong ECD life. Sending the ECD back is expensive and requires days of GC downtime, thus it should only be done when necessary. Cleaning the ECD should be attempted before returning it for reconditioning. Even if the detector is several years old, a thorough cleaning may extend its usable lifetime by 3-12 months. A hydrogen bakeout is the most common method and is suitable for all ^{63}Ni ECD's. Hydrogen is supplied to the detector cell either as the carrier gas (at a high flow rate) or as a substitute for the auxiliary gas. The column does not have to be removed prior to a hydrogen cell bakeout. Heat the detector to 350°C and leave the detector at this temperature for about 24 hours. Do not leave the detector in this configuration for more than 24 hours. After the hydrogen bakeout, return the GC to its normal operating state. After several hours, the ECD can be evaluated to determine the success of the bakeout. A more drastic cleaning method may be required if ECD performance has not returned.

The steam cleaning method appears to contradict some of the general guidelines about ECD's. There is disagreement between individuals whether this method should be used. There are numerous cases of very successful implementation of steam cleaning ECD cells. If there is any doubt, the GC manufacturer should be contacted before proceeding. A hydrogen bakeout should be attempted before a steam cleaning is undertaken. The procedure is simple, but it requires 1-2 days before the ECD is suitable for use. Remove the capillary column from the GC and install a short length of deactivated fused silica tubing or old column. Establish a low rate of carrier gas flow through the tubing. Adjust the ECD temperature to 350°C while maintaining the usual detector gas flows. Set the injector temperature above 200°C and the oven temperature above 150°C. Make 50-100 injections of 10-50 µL of deionized water into the GC. An autosampler is ideal for this procedure. The vaporized water steam cleans the ECD cell. No radioactive materials are removed from the cell. The ECD requires about one day before its background signal stabilizes. Steam cleaning followed by the hydrogen bakeout results in the cleanest detector. Several cleanings are usually possible before the detector has to be returned for reconditioning. The time interval between cleanings decreases as the detector approaches complete failure. Eventually detector performance no longer improves upon cleaning and reconditioning is necessary.

Most ECD's have a swab kit available from the manufacturer to check for leakage of source materials. Areas around the detector are wiped with a moist swab, sealed in a bag and sent to the GC manufacturer. The swabs are checked for radioactivity to make sure the ECD cell is not leaking source materials. This is a rare occurrence, but should be occasionally checked. If excessively high levels are found, the GC manufacturer will advise on the proper methods to handle the situation.

8.5.8   Common Problems
8.5.8.1   Change in Sensitivity

ECD response changes with temperature. The effect of a temperature change on the response depends on the compound, the original detector temperature and the direction of the temperature change. Most ECD's are very highly affected by the total flow of gases through the cell. Even a small change in the auxiliary or makeup gas flows causes a noticeable shift in the sensitivity. Changes in the carrier gas flow have a smaller impact. The affect of a flow change on response depends on the compound, the original flow rate, which gas has changed, the original flow of each gas, and the direction of the flow change.

A change in the amount of oxygen or water in the gases (carrier, auxiliary or makeup) shifts ECD response. In general, increases in water and oxygen decreases the ECD response; however, under specific conditions oxygen enhances the response of some compounds. Water and oxygen present at levels as low as 10 ppm affects detector sensitivity. Water and oxygen traps are usually installed on the detector gas lines.

As an ECD cell becomes dirty, response (i.e., peak size) may increase. This increase in response is always accompanied by an equal or greater increase in noise. The signal to noise ratio has not increased; therefore, the actual sensitivity of the detector has not improved and may have actually decreased. The response for each compound may not be affected equally by the condition of the cell.

8.5.8.2   Peak Shape Problem Attributed to the Detector

Improper column installation or too low makeup gas flow are the most common source of broad or tailing peaks caused by the ECD. If the column is positioned too low (i.e., below the cell entrance), there is excessive dead volume. If a column is forced too far into the detector, a broken column end may result. It is possible to insert the column into the cell for some ECD's. Having the column a small distance into the cell usually does not cause any peak shape problems; however, a sensitivity problem may occur. For all of these column position problems, tailing or broad peaks may result, with the problem being more severe for very volatile compounds.

Too low or the lack of makeup or auxiliary gas flows may cause tailing or broad peaks. ECD cells have large volumes, and thus require high makeup or auxiliary gas flows to minimize dead volume problems. Usually a large change in the sensitivity occurs with excessively low gas flows. Depending on the original gas flow, no change or a gain in the sensitivity is observed with the loss of peak shape.

8.5.8.3 Baseline Problems

A small amount of baseline drift may be evident as the carrier gas flow rate decreases during the course of a temperature program. The flow through the ECD cell changes, and the baseline therefore drifts corresponding to the temperature program. If the stationary phase contains a species that elicits an ECD response, normal column bleed is evident as a baseline rise. The rise should only occur as the temperature approaches the upper temperature limit of the column. Detector temperature fluctuations results in baseline drift as the temperature changes. A malfunction in the detector heating unit or sensor is the primary cause of a consistent temperature drift.

Water in the injected samples can cause appreciable noise problems which may last for some time due to adsorption effects within the detector. The baseline problem occurs around the solvent front or in the retention time region where water elutes from the column. The water peak is very broad and irregular for many columns with nonpolar ones being the worse. Depending on the detector, sensitivity settings and the amount of water, the disturbance may last only 1-2 minutes or up to 5-10 minutes. A large amount of a halogenated compound may also cause a substantial baseline disturbance.

8.5.8.4 Negative Peaks

In most cases, negative peaks are a sign of a highly contaminated detector or a detector that needs reconditioning. If the negative peaks appear after some of the positive peaks in the chromatogram, a dirty cell is the most likely culprit. Usually the size of the negative peak corresponds with the size of the positive peak. If the negative peaks are random or in the same location away from positive peaks, a contaminated cell may not be responsible.

There are electropositive compounds that can contribute electrons to the ECD cell. This increases the current in the cell, thus generating a negative signal. If the negative peak is always in the same location relative to the other peaks and its size corresponds to the amount of sample injected, there may be an electropositive compound in the sample. Compound causing negative peaks with an ECD are relatively rare. In some cases, a high concentration of a compound that does not produce an ECD response can affect the detector. The most common example are hydrocarbons. At very high levels, hydrocarbons in the sample can produce a quenching effect and are sometimes observed as negative peaks.

Negative peaks with an ECD are not unusual. A small negative peak immediately after the solvent front is fairly common. As long as the negative peaks do not interfere with any peaks of interest and do not seem to be related to the age or cleanliness of the ECD, they are not of major concern.

8.5.8.5　Linear Range Problems

It is quite easy to exceed the linear range of an ECD. The 10^3 range is relatively small, thus the potential concentrations of many samples fall outside the linear range. Dilution of the too concentrated samples and re-injection is sometimes necessary. Calibrations curves with a large number of points will also help to accurately quantitate samples in the nonlinear range of the response curve. Good calibration curves eliminate the need to dilute and re-inject samples.

One of the first signs of a contaminated (dirty) ECD cell is a decrease in the linear range. If frequent or large, multiple point calibration curves are not generated, the loss of linear range may not be noticed until the errors in the quantitative results become quite severe. As ECD cells become older, the linear range starts to decrease. Eventually, it will become too small to use for most applications.

8.5.8.6　Miscellaneous

Due to the sensitivity of ECD's to halogens, background contamination can be a problem. High volume use of volatile halogenated solvents such as Freons or methylene chloride in the vicinity of the GC can be detected as higher backgrounds. Sometimes the ECD background signal shifts, depending on the frequency of use, time of the last usage, and the volume used of the solvent. The airborne solvent vapors may contaminate vials, pipets, solvents, or anything else coming in contact with the samples. The high sensitivity of the ECD can respond to even very low levels of contamination. Depending on the column and GC operating conditions, a peak representing the contaminating compound is seen.

8.6 Thermal Conductivity Detector (TCD)
8.6.1 Principle of Operation

A simplified drawing of a generic TCD is shown in Figure 8-6. The detector consists of two heated cells. The carrier gas from the column flows through the sample cell; the reference cell is supplied with the same type and flow of gas that flows through the column. For most TCD's, each cell contains two interconnected filaments. The same amount of a controlled current is applied to each filament. Each filament establishes an equilibrium temperature if the current and gas flow rate across the filament remain constant. Gas flowing across the filament cools it to some extent by absorbing some of the heat. The amount of heat loss is a function of the thermal conductivity of the gas flowing through the cell. The ability of the gas to conduct heat away from the filament changes if the composition of the gas changes. The gas flowing through the reference cell does not change, thus the filament temperature remains constant. The gas flowing through the sample cell can change since it may contain compounds eluting from the column. If compounds pass through the sample cell, its filament changes in temperature since the composition of the gas changes. The resulting change in the sample cell filament temperature is compared to that of the reference cell filament temperature. The difference between the filament temperatures is converted into an electronic signal. This signal is processed by the recorder into a corresponding peak in the chromatogram. Greater amounts of the compound generate a larger signal, thus giving a larger peak.

Most TCD's were designed for use with packed columns. Their large cell volumes are not well suited for use with capillary columns. Large volumes of makeup gas are necessary, which reduces the sensitivity. A newer TCD designed for capillary columns utilizes a single cell and filament. The gas flow to the cell is very rapidly switched between the column effluent and reference gas. While the carrier gas from the column is flowing through the cell, it acts as the sample cell; while the reference gas is flowing through the cell, it acts as the reference cell. The difference in the sample and reference periods are measured and compared to generate the final signal. This design allows the use of a very small cell volume, and thus lower total gas flows. This increases sensitivity over the much larger cell volume TCD's.

Figure 8-6:   Major Components of a Dual Cell TCD

8.6.2 Gases

Makeup gas is usually required for optimal performance even with low volume cell TCD's. Gas flow rates are critical to the sensitivity of a TCD. The best sensitivity occurs at lower flow rates. Excellent results are obtained when the total gas flow is maintained at a minimum of about 5 mL/min. This ensures a rapid sweeping of the outside of the column and eliminates any dead volume problems in the detector. For most TCD's, flow rates below 5 mL/min result in tailing or broad peaks. Usually, a total flow rate of 5-20 mL/min for the carrier plus the makeup gas is used.

Hydrogen carrier gas provides the best sensitivities; however, its use may result in sensitivity changes as the filament ages. There is also the possibility of reactivity problems of the compounds with the hot filament in a hydrogen environment. Helium is the recommended carrier and makeup gas; it has sensitivities close to hydrogen without the possible reactivity problems. Nitrogen gives lower responses, but is useful if helium or hydrogen is being analyzed. A split peak may be obtained when trying to analyze hydrogen using helium as the carrier gas. A special mixture of 8.5% hydrogen in helium is recommended for analyzing hydrogen and minimizes the nonlinear response of TCD's at low levels of hydrogen.

8.6.3 Column Position

The position of the column relative to the detector cell is important for optimal sensitivity and peak shape. The end of the capillary column must be at the entrance of the sample cell. Placement of the column end in the cell results in poor peak shapes and poor sensitivities. Placement of the column end too far below the cell entrance may result in broad and tailing peaks (dead volume) and/or the loss (absorption) of active sample components. The GC's instruction manual is the best source for the proper positioning of the column in the detector.

8.6.4 Temperature

The temperature of the detector is very important. Sensitivity increases as the temperature difference between the cell and the filament increases. It is advantageous to keep the cell temperature as low as possible. It must be hot enough to minimize temperature fluctuations from the outside air and to prevent condensation of high boiling sample components. Temperatures in the range of 150-250°C are normally used. Lower filament currents require the use of lower detector temperatures to maintain good sensitivity.

8.6.5 Selectivity

TCD's are universal in their response. Any compound in the carrier gas elicits a response from a TCD. For this reason, the TCD is used in applications where a FID is not suitable. Analysis of permanent gases (O_2, H_2, N_2, He, Ar, Xe, etc.), oxides of nitrogen (NO, NO_2, N_2O, N_2O_2), sulfur compounds (H_2S, SO_2, CS_2, COS, etc.), CO_2, CO, and water are typical applications for a TCD.

8.6.6 Sensitivity and Linear Range

The minimum detectable quantity is 5-50 ng of compound. The primary drawback of a TCD is its lower sensitivity. The FID has about 10-100 times greater overall sensitivity, a greater linear range and more reliable signal for quantitation purposes. For these reasons, the FID is a more popular detector. The TCD responds to compounds that are not detectable with a FID (water, CO_2, CO, some sulfur compounds, oxides of nitrogen, and permanent gases). The C_1-C_3 hydrocarbons may give a better response with a TCD, and especially the low cell volume designs.

Sensitivity is increased as the temperature difference between the filament and the cell becomes larger. There are two methods to control the temperature difference. The easiest and most reliable method is to increase the filament current. The additional current increases the temperature of the filament. If the current is too high, noise may develop and there is the risk of premature burning out of the filament. The second method is to decrease the temperature of the cell (detector). The temperature must be hot enough to minimize temperature fluctuations caused by changes in the outside air. Also, the temperature should not be so low that high boiling sample components condense in the detector. Usually the filament current is altered since it is the easiest and quickest method to alter the temperature difference between the filament and cell. Experimentation with the amount of current to maximize sensitivity is suggested, but sensitivity criteria must be balanced against reduced filament lifetime and high noise at high current levels. The GC instruction manual provided some current settings as a starting point.

The TCD is a concentration dependent detector meaning sensitivity decreased at higher flow rates of carrier and makeup gases. Lower gas flows are preferred to maximize sensitivity, but the flow needs to be high enough to effectively sweep the cell. In many cases, optimal sensitivity and minimal peak broadening can not be achieved at the same flow rates; therefore, a compromise must be reached. Higher flow rates that sacrifice some sensitivity are used to maintain narrow peak widths. Another cause of reduced sensitivity is the lower filament temperature caused by high gas flows. Since the filament is hotter than the cell, this reduces the temperature difference between the filament and cell. This smaller temperature differences reduces sensitivity.

The linear range of most TCD's is about 10^5. This is similar to many detectors, but smaller than the FID. The TCD may be better for some sulfur compound analyses than some sulfur selective detectors. Specifically, the FPD (Section 9.7) is very selective and sensitive for sulfur compounds, but it is not linear in the sulfur mode. When samples are over a wide concentration range and detectable by a TCD, the TCD may be a better detector. The nonlinear response of the FPD makes quantitation difficult and requires the use of many calibration standards.

8.6.7 Maintenance

The primary maintenance for a TCD involves the filament. Most procedures involve improving filament life or keep the filament from becoming damaged or contaminated. A constant presence of oxygen can permanently damage filaments through oxidative processes. The most common source of oxygen is elevated levels in the carrier or makeup gas or a leak near the detector. Oxygen traps are recommended for the carrier and makeup gases to reduce oxygen levels. Proper column installation techniques and regular leak checks (especially after column installation) help to keep leak problems to a minimum. The damage caused by oxygen is more severe at high filament currents. Chemically active sample components such as acids and halogenated compounds may attack and damage the filaments. Avoiding these compounds when possible increases filament life. Turning off or substantially reducing the filament current when the TCD is not in use prolongs filament life.

Increased filament lifetime will result if the following start up scheme is used. Purge the detector with carrier and makeup gas for 10-15 minutes before turning on the filaments. This prevents oxidation of the filaments due to the presence of oxygen (air) that has diffused into the cell under no flow conditions. Turn on the filaments at the lowest possible current setting then increase the filament current in several increments to the desired value. This reduces the power surge upon current introduction and prolongs filament lifetime.

Cell contamination is a problem when lower a detector temperature is used to improve sensitivity. Also, low filament currents promote contamination since the filament is maintained at a lower temperature at lower currents. If the cell becomes contaminated, a solvent flush of the detector may help to remove the condensed material. Cool the cell to room temperature and remove the column. Place a septum in a nut or fitting assembly that fits onto the detector entrance. Place the nut or assembly on the detector fitting and tighten. Verify the presence of makeup gas flow. Inject 20-100 µL volumes of toluene or benzene into the detector through the septum. Inject a total volume of at least 1 mL of solvent. Do not inject halogenated solvents such as methylene chloride and chloroform into the detector. After the final injection, allow makeup gas to flow for 10 minutes or more. Slowly raise the temperature of the cell to 20-30°C above the normal operating temperature. After 30 minutes, decrease the temperature to the normal value and install the column as usual.

8.6.8 Common Problems
8.6.8.1   Change in Sensitivity

Sensitivity changes can usually be attributed to the filament. Any event or situation that changes the temperature of the filament affects sensitivity. On the most basic level, a change in the current setting changes the sensitivity (higher current = higher sensitivity). If the filament has be altered, a different current may be necessary to maintain a specific sensitivity. As filaments become older, the current necessary to obtain a specific sensitivity will change. Contaminated or damaged filaments also affect sensitivity. Contamination is a greater problem with low filament currents and detector temperatures. Obviously, injection of dirty samples or ones containing high molecular weight compounds will contaminate the cell or filament much quicker. High oxygen levels or samples containing high levels of halogenated or acidic compounds damage the filament.

Temperature and flow changes also affect sensitivity. TCD response changes with detector temperature. A lower detector temperature increases sensitivity. Changes in the makeup gas flows cause a noticeable shift in the sensitivity. Changes in the carrier gas flow have a smaller impact. Lower flows increase sensitivity, but broader peaks are usually evident at lower flow rates.

Leaks in the area of the detector can cause flow fluctuations and temperature variations. The change in the temperature and/or flow alters sensitivity. Excessive baseline noise or short filament life usually accompanies a leak in the detector.

8.6.8.2   Peak Shape Problem Attributed to the Detector

Improper column installation or too low makeup gas flow are the most common source of broad or tailing peaks caused by the TCD. If the column is positioned too low (i.e., below the cell entrance), there is excessive dead volume. If a column is forced too far into the detector, a broken column end may result. It is possible to insert the column into the cell for some TCD's. Having the column a small distance into the cell usually does not cause any peak shape problems; however, a sensitivity problem may occur. For all of these column position problems, tailing or broad peaks may result with the problem being more severe for very volatile compounds.

Too low or the lack of makeup gas flows may cause tailing or broad peaks. Most TCD cells have large volumes, and thus require high makeup gas flows to minimize the dead volume. Usually a large increase in sensitivity occurs with excessively low gas flows.

8.6.8.3   Baseline Problems

A small amount of baseline drift may be evident as the carrier gas flow rate decreases during the course of a temperature program. The flow through the TCD cell changes, and the baseline usually drifts corresponding to the temperature program. Depending on the sensitivity setting and the column, normal column bleed is evident as a baseline rise. The rise should only occur as the temperature approaches the upper temperature limit of the column. Detector temperature fluctuations result in baseline drift as the temperature changes. A malfunction in the detector heating unit or sensor is the primary cause of an inconsistent temperature drift.

Another cause of baseline drift may be the condensation of high boiling compounds on the filaments. This is a greater problem for samples containing high molecular weight compounds or when using low detector temperatures or filament currents. Temporarily increasing the filament current to volatilize the contaminants may help; however, excessive times at high currents decrease filament life.

If the baseline suddenly goes off scale and the signal is high, regardless of the instrument and recorder sensitivity settings, the filament is probably burnt out and needs to be replaced. If the signal gradually drifts too high, a burnt out filament is not the problem. A contamination, temperature or flow problem is the probable cause.

8.6.8.4   Negative Peaks

When mixed with the carrier gas, some compounds decrease the thermal conductivity of the gas flowing through the TCD cell. This increases the heat of the filament and result in a negative signal. A negative instead of positive peak occurs. Negative peaks for a TCD is not unusual. If all of the peaks are negative, check the polarity of the detector and recorder to ensure proper connection.

8.6.8.5   Short Filament Lifetimes

Short filament life is normally caused by excessively high filament currents. Decreasing the current prolongs filament life, but decreases sensitivity. If very high filament currents are necessary to detect the sample compounds, placing more sample on the column (if possible) or switching to a more sensitive detector may be necessary.

Oxygen damages TCD filaments especially at high filament currents. A leak in the GC system is the most common source of oxygen. Acids and halogenated compounds may attack and damage the filaments. Using hydrogen as the carrier gas usually accelerates chemical or oxygen damage.

8.7   Flame Photometric Detector (FPD)
8.7.1 Principle of Operation

A simplified drawing of a FPD is shown in Figure 8-7. Carrier gas exiting the column is mixed with oxygen then introduced into a hydrogen flame via a jet. Compounds in the carrier gas are burned in the flame. The sulfur compounds form S_2 and phosphorus compounds form HPO. Under the conditions of the combustion flame, S_2 emits light with a maximum around 394 nm and HPO at a maximum of around 526 nm. In the sulfur mode, a filter is used to screen out all wavelengths except 394 nm. In the phosphorus mode, a filter is used to screen out all wavelengths except 526 nm. Only the specific wavelength corresponding to the sulfur or phosphorus species reaches the photomultiplier tube (PMT). Some FPD's contain two filters for the simultaneous monitoring of both wavelengths. Upon stimulation by the transmitted light, a current is generated by the PMT. This signal is processed by the recorder into a corresponding peak in the chromatogram. Greater amounts of the compound generates a larger signal, thus giving a larger peak.

Figure 8-7: Major Components of a FPD

There are dual burner FPD's that use two flames. The first flame is for decomposing the compounds and the second flame for formation of the light emitting species. Two different air supplies (flows) are used to maintain the flames. This designs minimizes the quenching problem of FPD's (Section 8.7.6). Also, this design prevents large amounts of solvent from extinguishing the flame (solvent flameout) that occurs with some single burner FPD's. The filter and PMT portions of a dual burner FPD are the same as a single burner design.

8.7.2 Gases

Hydrogen and air are used as the combustion gases. There are different detector gas flows for the sulfur and phosphorus modes. The actual flows are found in the GC instruction manual. They may vary between the different manufacturers due to differences in the designs. For the sulfur mode, hydrogen flows of 60-80 mL/min and air flows of 80-100 mL/min are common. The flows are usually set so that the ratio of oxygen (in the air) to hydrogen is 0.2-0.3. For the phosphorus mode, hydrogen flows of 120-170 mL/min and air flows of 100-150 mL/min are common. Sensitivity changes may result if the hydrogen and air ratios are changed.

Makeup gas of nitrogen enhances the sensitivity of the FPD especially in the sulfur mode. The nitrogen gas flow should be 50-100 mL/min, but this value is not critical. Actual flow values can be found in the GC's instruction manual. Makeup gas flow rate affects sensitivity, thus a correct and consistent flow is important for the best performance.

The choice of carrier gas is not critical. Using hydrogen or nitrogen as the carrier for wide diameter columns may result in slight response changes or baseline drift during a temperature program. Usually the changes are not noticeable.

8.7.3 Column Position

The position of the column within the detector is critical for FPD's. Most sulfur and phosphorus containing compounds are very active and are readily absorbed by non-inert surfaces. The end of the column should be at the tip of the jet to minimize dead volume and compound contact with active surfaces. If the column end is below the tip of the jet, broad or tailing peaks may result due to dead volume effects. Adsorption of active compounds may also occur with tailing or reduced size peaks. It is easy with some FPD's to insert the column into the flame. The column end will burn in the flame and excessive noise and baseline problems occur.

8.7.4 Temperature

The PMT is positioned close to the burner assembly. This limits the detector temperature to 150-275°C. PMT noise increases with increasing detector temperature and sensitivity decreases with increasing detector temperature especially in the sulfur mode. For these reasons, lower detector temperatures are normally used. High molecular weight compounds may condense in the detector at the lower temperatures, thus creating a problem with detector contamination. Areas such as the filter are particularly susceptible to contamination and have a significant impact on sensitivity. A balance between sensitivity and detector contamination has to be reached; however, it only becomes an issue when analyzing higher molecular weight compounds. Some FPD's have separate temperature control of the detector body and base. The temperature of the detector base is less critical. Even for the analysis of very volatile compounds, a detector body temperatures of at least 125°C are recommended so that water condensation does not occur.

8.7.5 Selectivity

The selectivity of a FPD for sulfur and phosphorus over hydrocarbons is 10^4-10^5. Due to the overlap in the emission wavelengths, the selectivity of phosphorus over sulfur is in the range of 5-10. This results in the possible detection of high concentrations of sulfur compounds in the phosphorus mode. The selectivity of sulfur over phosphorus is 10^3-10^4, thus there is practically no direct interference from phosphorus compounds in the sulfur mode.

8.7.6 Sensitivity and Linear Range

The minimum detectable quantity is 10-100 pg of sulfur compounds and 1-10 pg for phosphorus compounds. Increased sensitivity occurs at lower detector temperatures; however, contamination of the detector increases at lower temperatures. Changes to the hydrogen, air or makeup gas flow rates affect the sensitivity. The gas, amount and direction of the flow change, and the compound determine the amount of the sensitivity shift. If hydrogen is used as the carrier gas, a large change in its flow affects sensitivity. Hydrogen flow to the flame changes, thus the sensitivity also changes.

The presence of non-sulfur and non-phosphorus compounds in the sample can present a problem. If one of these compound even partially co-elutes with a sulfur or phosphorus compound, the response of the compound can be reduced. This effect is called quenching. While any compound can cause quenching, hydrocarbons are particularly effective at reducing FPD response. Sulfur compounds seem to be more affected than phosphorus compounds. Usually a 10-50 fold excess of the quenching compound significantly reduces the response. The greater the concentration, the greater the amount of quenching. In many cases, there is a complete loss of FPD response with complete co-elution of a quenching compound. The linear range of a FPD for phosphorus is 10^3-10^5. FPD responses are nonlinear for sulfur. Values up to 10^2-10^3 sometimes are reported, but this range is very dependent on the operating conditions, detector design and especially the compound. Calibration curves for FPD's are usually obtained by plotting two times the log of the response versus the log of the sulfur concentration. Errors can still be introduced with this approach. In any case, multiple point calibration curves over the entire possible concentration range of each compound are needed.

8.7.7 Verifying Flame Ignition

The signal output as displayed on the data signal or GC display can be used to verify flame ignition. Check the signal prior to lighting the FPD. The signal should be higher when the flame is burning. By comparing the two values, it is easy to determine if the flame is lit. The actual numerical values depend on the source of the detector signal. Different GC's and data systems report detector output in different units or values. This is not a problem since only a comparison between the lit and unlit detectors are needed.

To physically verify that the flame is lit, place a cool, shiny object (e.g. wrench, knife blade, etc.) over the vent of the detector. After several seconds, condensation should be visible on the shiny object after it is removed from the vent. If no condensation is visible, the detector flame may not be lit.

8.7.8 Maintenance

FPD's require minimal maintenance to keep them performing at satisfactory levels. The hydrogen, air and makeup gas flows should be occasionally measured. They can drift over time or be unintentionally changed without knowledge of it occurring. Each gas flow should be measured independently to obtain the most accurate values.

The FPD requires periodic cleaning. In most cases, this only involves the jet and less frequently, the filter. Cleaning kits are available containing brushes and wires that simplify the cleaning of all of the detector parts. The brushes are used to dislodge particulates clinging the metal surfaces. A fine wire is used to clean the jet opening of particles. Do not force too large a wire or probe into the jet opening or the opening may become distorted. A loss of sensitivity, poor peak shape and/or lighting difficulties may result if the opening is deformed. The filter or any of the window parts should be handled gently. Scratches or other surface deformities reduce the amount of light passing through the filter, thus reducing response. The filter and related parts should be clean and free from fingerprints.

The PMT need periodic replacement. High detector temperatures reduces the PMT life. When not in use, turn off the PMT to maximize its usable lifetime. Some PMT's may have a shelf life and should not be stored for prolonged periods before use.

8.7.9   Common Problems
8.7.9.1   Change in Sensitivity

A change in detector temperature affects sensitivity. FPD response decreases with an increase in the detector temperature. Sensitivity is dependent on the flow rates of hydrogen, air and the makeup gas. Changes in the makeup gas flows cause a noticeable shift in the sensitivity; changes in the carrier gas flow have a smaller impact. Lower makeup gas flows increase sensitivity, but broader peaks are usually evident at lower flow rates. Leaks in the detector can cause flow fluctuations and temperature variations. The change in the temperature and/or flow alters sensitivity.

Quenching is another cause of sensitivity problems. FPD response can be severely reduced if a non-sulfur or non-phosphorus compound partially or fully co-elutes with the analyte of interest. Eliminating or reducing the co-elution increases the peak response. Changing the temperature program or column may remove the co-elution, thus drastically improving sensitivity.

Contamination of the filter window reduces FPD sensitivity. Cleaning or replacement of the window is necessary to correct this problem. As the PMT ages, sensitivity decreases either as reduced peak size or increased noise.

8.7.9.2   Peak Shape Problem Attributed to the Detector

Improper column installation in the detector or too low makeup gas flows are the most common source of broad or tailing peaks caused by the detector. If a column is forced too far into the detector, a broken column end may result. The broken column end and resulting debris disrupt the flow at the end of the column. This turbulent flow may cause broad or tailing peaks.

When a column is too far away from the tip of the jet, there will be excessive dead volume. Tailing or broad peaks may result with the problem being more severe for very volatile compounds. Too low or nonexistent makeup gas flows often cause tailing or broad peaks. Significant sensitivity loss usually accompanies insufficient makeup gas flows.

8.7.9.3   Loss of Linear Range

The most common cause of linear range problems is a change in the detector gases with the makeup gas being the most influential. The linear range in the sulfur mode is very small. It depends on the operating conditions and the compound. Multiple point calibration curves are essential for FPD's.

8.7.9.4   Flame Goes Out

Large sample volumes (usually above 2 µL) may extinguish the detector flame. This usually occurs when the sample solvent passes through the detector. There is little that can be done to eliminate the problem except to inject a smaller volume. Probably the largest factor concerning solvent flameout is the design of the FPD.

8.7.9.5   Miscellaneous

Some FPD's pop loudly upon flame ignition and this is normal. This is due to the high hydrogen flow rates that are used and the large volume of FPD's. A buildup of hydrogen in the detector occurs very rapidly at these high flow rates. Ignition of the flame immediately after the hydrogen gas is turned on reduces the frequency and severity of the popping.

8.8   Mass Spectrometers (MS)
8.8.1 Principle of Operation

A simplified drawing of a representative quadrupole MS typically used as a GC detector is shown in Figure 8-8. The entire mass spectrometer is maintained under a vacuum of 10^{-4}-10^{-6} torr by use of diffusion or turbomolecular pumps. Carrier gas exiting from the column is introduced into the ionization chamber or source via an interface. The source is heated to 150-300°C to prevent condensation of the sample. A filament outside of

Figure 8-8: Major Components of a Quadrupole Mass Spectrometer (Not to Scale)

the source is used to produce a supply of electrons. The electrons enter the source through a small hole near the filament. A negatively charged shield behind the filament directs the electrons towards the hole in the source. Magnets are often used to focus the electrons into a narrow beam. The potential difference between the filament and source controls the energy of the electrons entering the source. Larger potential differences increases the energy of the electrons. A potential difference that imparts an electron energy of 70eV is used in most cases. When the electrons enter the source, some of them collide with the molecules eluting from the column. The collision imparts energy throughout the molecule which then becomes ionized. If a sufficient amount of energy accumulates in a single bond or group of bonds, the molecule dissociates into smaller, positive ion fragments. The fragments produced are primarily dependent on the structure of the original (parent) molecule and the energy of the electrons. The positively charged fragment ions are ejected from the source by being repulsed by a repeller held at a slightly positive potential. The ions travel through a series of focusing plates which accelerate the ions and focus them in a narrow beam. The focussed beam of ion fragments then enters the quadrupole analyzer. The quadrupole is four accurately machined rods set in a square array. Inside of the array a complex field is generated by potentials applied to the rods. The ion fragments entering the quadrupole are set into sinusoidal oscillations. At a particular field, only ions with a specific mass-to-charge (m/z) ratio can pass completely through the quadrupole array without colliding with one of the rods. Any ion that collides with the rods does not reach the detector at the end of the quadrupole array. The field is continuously swept at a very fast rate (usually 2-4 scans per second) to allow ions within a range of m/z ratios to pass through the quadrupole. This allows the quadrupole to act as a mass filter. The detector is at a negative potential so to attract the ions from the quadrupole. Most GC/MS systems use an electron multiplier as a detector which acts to amplify the signal resulting from the impact of the ions. As each ion impacts the detector, its m/z ratio is extrapolated from the value of the potentials applied to the quadrupole rods. The amount of each ion and its m/z ratio measured during one scan is stored in the data system. All of the scans during a run are sequentially combined. The total intensity (abundance) of the ions detected during one scan is plotted versus the start time of the scan to produce a total ion chromatogram (TIC). A background or baseline signal is always present due to the presence of the carrier gas impurities, system contaminants and normal column bleed. Introduction of a compound into the source produces a higher level of ions, thus giving a detector signal above the background. A peak is generated whose size corresponds to the amount of the compound.

8.8.2 Mass Spectral Data

Figure 8-9 shows a TIC. Retention time and peak area data are obtained from a TIC in the same manner as any other chromatogram. The additional information available is the m/z ratio and intensity (abundance) of all the ions detected during one scan anywhere in the chromatogram. The plot of this data is called a mass spectrum (Figure 8-9). The mass spectrum obtained from the peak is for the compound represented by that peak. Each line in the spectrum represents a fragment of the parent molecule. In many cases, the parent molecule does not completely fragment and retains its original structure. This ion is called the molecular ion.

Figure 8-9:   TIC and Mass Spectrum

The fragmentation pattern of a given molecule is the same each time providing the electron energy used for fragmentation is the same. The mass spectrum from a peak in the chromatogram can be compared to known mass spectra stored in the computer library or produced from the analysis of analytical standards. If the patterns match, the identity of the compound is confirmed with a high degree of certainty. Along with the retention time data, very high confidence identification of compounds is possible.

The mass spectrum from a single peak representing two co-eluting compounds is a composite of the two spectra. The identification of compounds from their mass spectra is difficult and uncertain for co-eluting peaks. Fortunately, good mass spectra can be obtained from partially co-eluting peaks. The mass spectrum for each compound is obtained from the portion of the peaks that do not overlap.

The practical mass limit for most quadrupole GC/MS systems is 600-800 amu (mass units). Many systems allow scanning at the higher mass ranges, but a loss of many of the higher mass fragments occurs. One of the features of the more expensive quadrupole systems is their higher mass range.

8.8.3 Other Ionization, Detection and Mass Filtering Modes

Ionization and fragmentation of molecules using electrons is called election impact or ionization (EI). Another ionization technique is called chemical ionization (CI). Instead of electrons, small gas molecules are used. Methane, butane and ammonia are the most common gases used for CI. CI is considered a softer fragmentation technique. Less fragmentation occurs, thus fewer fragments are produced. Some molecules undergo massive fragmentation with EI and no molecular ions are formed. The large number of low abundance fragments and the lack of a molecular ion makes identification more difficult and uncertain. Some isomeric molecules have very similar EI spectra, but the CI spectra are unique allowing unequivocal identification. Most GC/MS systems require some amount of dismantling and downtime to switch between the two ionization modes. CI is usually an option on most systems, but at additional cost. Coupled with the fact that EI works for most compounds and is simpler, EI is the most common ionization mode.

Instead of an electron multiplier, a few GC/MS systems use a conversion dynode and a photomultiplier tube (PMT). The ions impact the diode and generate secondary electrons that collide with a phosphor disk. This creates photons which are detected and amplified by the PMT. The number of photons hitting the PMT is directly related to the number of ions leaving the quadrupole. PMT's have some advantages over electron multipliers as MS detectors. In general, they provide more sensitivity and much longer lifetimes.

Another design of a benchtop MS system is called the ion trap detector (ITD). The primary difference is the mass filter portion of the unit. Instead of the ions travelling in a linear pattern (i.e., through the quadrupole array), the ions in an ion trap travel in a circle. Ions with the incorrect m/z ratio are ejected from the circular pattern and do not reach the detector. The ion trap collects and detects more of the generated ions than a quadrupole. This makes ion traps about 10-20 times more sensitive than quadrupole GC/MS systems. The higher cost and complexity, sometimes unique EI spectra and concentration dependent spectra makes ion traps less common than quadrupole based GC/MS systems.

8.8.4 Selectivity

MS selectivity is based on the mass of the fragments. The masses detected and measured are selected by the analyst. If the mass spectrometer scans for all the masses over a broad range, this is called the full scan mode. All of the ions that impact the detector are collected. A full mass spectrum is available for each peak. If the most abundant and unique mass fragments of a compound are known, selected ion monitoring (SIM) can be used. The SIM mode does not scan all of the masses, but only a select few. Only the selected ions are collected and all of the others are ignored. Only the compounds that have one or more of the selected mass fragments are represented by a peak in the chromatogram. Since only a few ions are monitored, obtaining a full mass spectrum or information on non-selected ions is not possible. Most systems allow the selected ions to be changed at preselected times during the run. This improves selectivity by limiting response only to the ions expected for the peaks eluting within a particular retention time region.

Usually 2-3 of the most abundant or unique ions for a compound are monitored in the SIM mode. If the masses are carefully chosen, very few compounds have the same mass fragments especially at the same relative abundances. For a known set of compounds, the identity of each compound can often be determined by the presence and abundance of its unique set of ions. Co-eluting peaks present less of a problem in the SIM mode providing each compound has an unique set of ions. By measuring only the ions unique to each compound, the presence and amount of each compound can be determined.

The SIM mode is often used to screen samples. If it appears that a certain compound is present based on the SIM mode information, the sample is re-run using the full scan mode. The complete mass spectrum is more definitive as a means of compound identification. Many isomers and homologs have the same major ions, but the fine detail available only in their full mass spectra can be used to distinguish which compound is actually present.

8.8.5 Sensitivity and Linear Range

MS sensitivity depends primarily on the scan mode. The sensitivity in the full scan mode is 0.1-1 ng of compound. The sensitivity in the SIM mode is 0.1-10 pg of each compound. Better sensitivity is obtained when the selected ions are substantially more abundant than the non-selected ions. More of these ions are created, thus boosting the number reaching the detector. Extensive fragmentation results in numerous ions, but all of them are at low abundances which lowers the intensity of their detector signal. If too many ions are selected in the SIM mode, the sensitivity advantages are reduced. For analyses with a large number of compounds, the selected ions are changed during the run to the ones corresponding to the peaks expected within a particular retention time range. This minimizes the number of ions selected at any one time and helps to maintain the high sensitivity available with SIM.

Better response is obtained if the electron multiplier voltage is increased. However, this also increases the noise, and thus no real increase of sensitivity is obtained. Higher voltages are required as the multiplier gets older. Eventually the multiplier needs replacing since the noise becomes too high.

Better sensitivity is obtained at lower carrier gas flow rates. Using lower linear velocities or smaller diameter columns reduces the amount of carrier gas reaching the source. Longer run times and the possibility of an efficiency loss occur when reducing the linear velocity. The major sacrifice with smaller diameter columns is a lower sample capacity.

The linear range of most quadrupole MS systems is 10^4-10^5. The linear range is primarily influenced by the electron multiplier and its operating voltage. The linear range will shrink as the multiplier ages. Linear range is often called the dynamic range for MS systems equipped with electron multipliers.

8.8.6 Temperature

The transfer line, source and quadrupole analyzer are maintained at higher temperatures. This is mainly to prevent compounds from condensing on their surfaces. The transfer line is heated to keep the sample rapidly moving towards the source. Some systems use the heat of the transfer line to conductively heat the source. The source temperature remains 100-150°C below the transfer line temperature. The transfer line and source temperatures can not be independently controlled, but this rarely causes a problem. Other systems use a separate source heater, thus permitting independent temperature control. Transfer line and source temperatures of 200-300°C are the most common. Source temperatures does not have a significant affect on sensitivity. It just needs to be high enough to prevent condensation of contaminants on the source. Most systems do not directly heat the quadrupole analyzer. The quadrupole temperature is normally not directly controllable and its temperature is a function of the source temperature.

Systems that use a single heater require longer times to equilibrate after a temperature change. The transfer of heat to the various regions requires more time to occur and stabilize than if the areas were directly heated. Longer equilibration times are needed after periods of non-heating such as during column installation and system maintenance. Excessive background noise, baseline drift and calibration (tuning) difficulties are the usual symptoms of an unequilibrated system.

8.8.7 Column Position

There are several different designs of the interface between the MS source and the capillary column. Regardless of the design, the column effluent is directed into the interior of the source. In some systems, the column is fed through a transfer line until it is positioned in or adjacent to the source. Other systems use a transfer line usually containing a length of deactivated fused silica tubing. The column is connected to the transfer line which directs the column effluent into the source. Either design directs the column effluent into the source.

A column or transfer line that is not properly positioned creates problems. A large amount of dead volume occurs if the column or line end is to far away from the source. Tailing or broad peaks may result due to the dead volume. A loss of peak size may occur due to adsorption of active compounds on hot metal or glass surfaces. It is possible with some interfaces to push the column or line too far into the source. Excessive noise, sensitivity loss or complete failure of the system usually occurs in these situations. Sometimes only a degradation of peak shape is seen.

8.8.8 Carrier Gas Flow Rate Considerations

Higher carrier gas flows reduce the sensitivity of MS systems. The carrier gas inhibits the ionization and fragmentation of the compounds. Larger volumes of carrier gas decrease the sensitivity by reducing the number of ions reaching the detector. The flow limit of most benchtop GC/MS systems is 1-2 mL/min. Higher flows can be used, but tremendous sensitivity losses occur.

One method to reduce carrier gas flow is to operate at lower linear velocities. A lower volume of gas reaches the source, but a loss of efficiency may occur and retention times will increase. The best method to reduce carrier gas flow rates is by using smaller diameter columns (0.18-0.25 mm I.D.). Small diameter columns operate at high linear velocities, but low flow rates. An added bonus is the higher efficiency of smaller diameter columns, but longer retention times and lower sample capacities also occur.

If large diameter columns and their high carrier gas flow rates are required with MS detection, a jet separator is usually needed. The column effluent passes through the jet separator before it enters the MS. The jet separator acts to remove the lower molecular weight compounds. Since the carrier gas is the lowest molecular weight compound in the column effluent, large amounts of the carrier gas are removed. A much lower amount of the heavier compounds are removed from the column effluent. This reduces the amount of carrier gas reaching the source without substantially reducing the amount of the compounds. A greater number of light compounds than heavier compounds are lost via the jet separator; thus the jet separator discriminates against the lower molecular weight compounds. Substantial losses of volatile compounds can occur when using a jet separator. Jet separators require a vacuum pump to function. The jet separator and pump adds thousands of dollars to the cost of the GC/MS system in addition to increased maintenance requirements.

8.8.9 Maintenance

The GC/MS instruction manual is an essential resource for maintenance schedules and procedures. It contains specific information about the exact model of GC/MS in use. The main areas that need routine maintenance or attention are the vacuum pumps, source, and the tuning parameters.

Most GC/MS system have a roughing or foreline pump. The oil level should be checked every 1-2 weeks. If it is low, fill the pump with the oil recommended by the manufacturer. It is recommended to change the oil about every 6 months. If there are any traps associated with the pump, the trap should be replaced or refilled whenever the pump oil is changed. The main pump on most GC/MS system is a diffusion or turbomolecular pump. If these have any oil or fluids, they need to be checked every 6-12 months. It is advisable to replace the oil or fluid every 12 months.

The ion source becomes fouled with use. High molecular weight materials from the samples and other sources slowly accumulate on the ion source. It requires occasional cleaning as necessary. The manual has precise instructions on removing, disassembly and cleaning the source. The instructions need to be followed exactly or performance problems and/or damage to the MS will occur. The general procedure for source cleaning is polishing of the metal parts with an abrasive slurry. The parts are then rinsed and ultrasonicated in a series of solvents. After thorough drying, the ion source is reassembled. Extra care should be taken to ensure none of the parts have any residues of solvent, abrasive materials, fingerprints, etc.. The frequency of source cleaning depends on the samples injected into the GC/MS. Dirtier samples increase the frequency of source cleaning.

Tuning is the process of adjusting the MS parameters to obtain the best performance. Tuning is preformed by introducing a specific compound with well understood properties into the MS. While the compound is present in the MS, various parameters are adjusted until predetermined performance criteria are met. The MS can be tuned for good performance over a wide mass range for general purpose use or over a limited mass range for specific purposes. The primary performance criteria for most tuning parameters are sensitivity, relative ion abundances and mass assignments. The tune compound and criteria are determined by the analyst, stated in a particular method or as recommended for the GC/MS system. Most GC/MS have a variety of built in tuning programs. They vary mainly in the amount of automation and tuning criteria. The frequency of tuning depends on a number of factors. Any time a major change is made in the system it will need tuning. A maximum of one week between tunings is usually suggested. Many methods require tuning procedures to be performed more frequently. There is no harm in tuning a MS too much except for the time wasted making unnecessary tuning adjustments.

The filaments and electron multiplier need occasional replacement. Filament life depends on its treatment. In general, having the filament on when large amounts of solvents or oxygen are in the source reduce its life. Electron multipliers have a finite life of 6-24 months. Dirty samples foul the multiplier and unnecessarily high voltage reduce its life.

8.8.10   Common Problems

Due to the greater complexity of mass spectrometers, there are more possible problems that can occur than with standard GC detectors. Also, there are problems unique to the MS detector. The instruction manual is the best source of troubleshooting information. Most of the data systems have diagnosis programs and provide error messages. All of these are useful when trying to diagnosis and solve problems.

8.8.10.1   Change in Sensitivity

The source of most sensitivity shifts is the electron multiplier. Either it is dirty or the voltage is too low. Increase the multiplier voltage to see of the sensitivity increases. If the sensitivity does not increase, look for another source of the problem. As electron multipliers age, they require more voltage to maintain a reasonable sensitivity. Eventually, they reach the maximum voltage allowed and can not be made more sensitive. The multipliers need to be replaced if they reach the maximum operating voltage. Since electron multipliers rarely look dirty, visual inspection can not be used to determine multiplier performance.

A dirty ion source reduces sensitivity. The residues or debris reduces the ionization efficiency. This decreases the number of ions reaching the detector, thus reducing sensitivity. The filament has to be cleaned using the procedure in the instruction manual. An old or damaged filament decreases sensitivity. Insufficient electrons are produced to ionize and fragment the compound molecules. Replacing the filament or switching to the other filament (if available) is required.

If the vacuum pressure is too high, poor sensitivity results. Usually a leak is the cause of excessively high vacuum pressure. Another cause of high vacuum pressure is when the carrier gas flow is too high. In either case, the number of ions reaching the detector is reduced, thus explaining the loss of sensitivity. Other possible causes of sensitivity changes are different source temperatures and an incorrect tune. Check both to determine whether they may a source of the sensitivity problem.

8.8.10.2   Excessive Noise or High Background

Excessive noise or high background can be due to an old electron multiplier. Higher voltages are needed to obtain the desired sensitivity. The higher voltages increase the noise levels. Usually the noise levels increase gradually as the multiplier ages. If the noise level suddenly increased or the multiplier is fairly new, look for other sources of the noise. Usually it will be from an air leak, contamination or a damaged or old filament.

Contamination of the MS or the GC can cause high background levels. Contaminants in the GC can be difficult to diagnose and identify. Most of them originate from the samples and can vary tremendously, thus the masses rarely help to identify the source of the contamination. Sometimes the masses can be used to help identify the possible contaminants. Table 8-2 lists some of the more common masses and sources. If the ion source was recently cleaned, it is common to see residual solvent masses especially if the parts were not thoroughly dried. Diffusion pump fluid usually enters the MS from a sudden or improper venting of the vacuum. Vapors of diffusion pump fluid can enter the MS if there is no carrier gas flow for prolonged periods. Hydrocarbons originate from fingerprints on ion source parts (poor cleaning technique) or foreline/roughing pump oil (foreline pump on with no

Table 8-2:   Common Contamination Masses

Substance	m/z
Solvents:	
Methanol	31
Acetone	43, 58
n-Hexane	43, 57, 71
Toluene	91, 92
Silicones	73, 147, 207, 221, 281, 295, 355, 429
Diffusion pump fluid	170, 262, 354, 446
Phthalate plasticizers	149
Hydrocarbons	masses spaced 14 apart (fingerprints, foreline or roughing pump oil)

carrier gas flow and the diffusion pump off). Silicone contamination can originate from many sources. The most common sources are the septum and silicon based stationary phases.

The MS adjusts voltages or currents to maintain the desired electron supply from the filament. As filaments age, they have more difficulty in producing a good supply of electrons. Excessive noise may occur when the filament becomes old or if it is damaged.

8.8.10.3 Leaks

Leaks are normally due to vacuum seals that are not properly seated or damaged. The column nut at the MS interface is one of the most common leak sites. The various seals and O-rings around the MS are the other probable locations. Leaks are evident from several different symptoms. The most common are higher than normal vacuum pressure, high background, poor sensitivity and masses characteristic of air which are m/z 18, 28, 32, 40 and 44 (water, nitrogen, oxygen, argon and carbon dioxide, respectively). The most reliable method is to check the abundances of these ions. The abundance of m/z 28 should be less than the abundance of m/z 18. If an air leak is present, the ratio of m/z 28 to m/z 32 is about 5:1.

One method to locate a leak is to use an inert gas such as argon. Obtain a small lecture bottle of argon and install a flow regulator. Set up the MS to look for m/z 40. Spray the argon over the areas where a leak may be possible. After spraying each area, wait for 10-20 seconds. If a sudden increase at m/z 40 occurs, the leak is located in the area where the argon was sprayed. Repair the leak and check the other areas in case there are multiple leaks. Tightening the appropriate fittings or replacing the seal or O-ring are the usual repair techniques for leaks.

9   COLUMN INSTALLATION

9.1   The Importance of a Properly Installed Column

The importance of a properly installed capillary column is under-appreciated in most cases. A poorly cut or installed column can lead to inferior chromatographic results. Also, permanent damage to the column may result if the proper installation precautions are not taken before column conditioning and use. Any gas chromatograph is limited by the weakest portion of the system - column installation is a part of the system just as much as the column or the instrument itself.

9.2   Installing Fused Silica Capillary Columns
9.2.1 Column Installation Steps

The basic column installation steps are the same for practically every column and GC system. Every precaution applies to all columns and situations. Some of the later steps are not necessary to properly install a column; however, they are verification of a proper installation and often prove to be very useful in the event that troubleshooting is necessary at a later date. The basic steps are:

1. Cutting the column
2. Installation in the injector
3. Turning on and verifying the carrier gas flow
4. Installation in the detector
5. Verifying column installation and detector operation
6. Conditioning the column
7. Setting the linear velocity
8. Performing a bleed test
9. Injecting a column test mixture

9.2.2 Cutting Fused Silica Capillary Columns

For the best performance, it is critical for both ends of a fused silica capillary column to be even and without any burrs, chips or uneven areas. This is accomplished by using the proper technique and tools when cutting the column. A carbide or diamond tipped pencil, a sapphire cleaving tool or ceramic wafer is needed to cut the column. Scissors, files or similar tools are not acceptable cutting tools. The choice of a cutting tool is primarily a personal preference and perhaps a cost issue. The cutting tool should only be used to cut columns and not other items or materials. Regardless of the tool, the basic technique is the same.

The tip of a pencil or the edge of a wafer or sapphire tool are used as the cutting edge. Lightly scribe the polyimide coating with the cutting edge. The goal is to create a weak spot in the polyimide coating that exposes the fused silica tubing. Attempting to cut through the fused silica tubing with the cutting tool results in an uneven and unsatisfactory cut since the column is crushed in this area. Place a short length of the column against a finger or palm. This makes it much easier to feel how much pressure is being applied to the tubing by the cutting tool. The tubing flexes if it is not supported which makes it difficult to control the amount of applied pressure. Lightly drag the cutting tool edge across the column in the desired location. Gasp the column 1-2 cm below the scribe mark. Gently bend the part of the column above the scribe away from the scribe mark. If properly scribed, the column breaks easily and evenly. If the column does not break, re-scribe the column about 1 cm below the previous scribe. Attempt to break the column again making sure it breaks at the point of the new scribe and not at the first one. A small amount of practice is required to consistently cut a fused silica column. After a while it becomes easy and routine; however, the cut should always be inspected regardless of one's experience and skill. Sometimes the column breaks unevenly due to no fault or error in the cutting technique or method.

It is recommended to inspect the column ends with a magnifier. After cutting, the ends of the column should be examined with a 10-20X magnifier to verify a clean and straight cut has been obtained. If the column end is jagged or uneven, re-cutting is necessary. The column needs to be re-cut until a satisfactory end is obtained. A poorly cut column may result in peak tailing or adsorption. Small pieces of the jagged edge of the column can fall into the column and cause severe peak tailing or blockage of the tubing. Small pieces may also enter the detector and cause it to malfunction. Regardless of the analysis, column or GC system, a good column cut is essential to obtain the best performance.

9.2.3 Column Placement in the GC Oven

Most GC's have a column hanger in the oven. Some hanger designs only allow one column to be placed on the hanger. If a second column is installed in the same GC oven, another approach has to be taken. Sometimes the second column can be hung directly on the cage of the first column. If this is not possible, there are usually several holes in the top of the GC oven where small pieces of wire or paper clips can be attached. The wire can be fashioned into a suitable hanger and the second column cage can be hung from the pieces of wire.

Unwind enough of the tubing on both ends of the column so there is ample tubing free from the cage. The column should be placed in the GC oven so that there is no contact between the tubing and other objects in the GC oven such as column identification tags, oven walls, column hangers, etc.. The forced air currents inside the oven cause movement of the column and these contact points may abrade the column. This may lead to breakage at these abrasion points since the polyimide has been weakened in these areas.

Avoid tight or sharp bends in the tubing. These sharp bends place an excessive amount of stress on the tubing, thus increasing the possibility of breakage. The avoidance of sharp bends becomes more critical as column diameter increases. Large diameter tubing can not tolerate sharp turns or bends as well as small diameter tubing; therefore, it is more susceptible to breakage due to stress. Regardless of tubing diameter, it is much better to remove several centimeters of column length instead of subjecting the column to an extreme bend or turn. The loss of several centimeters has no affect on column performance.

9.2.4 Column Installation in the Injector

After placing the column in the GC oven, the next step is to install the column in the injector. Place the column nut and ferrule on the tubing making sure of the proper alignment of each piece. Consult the GC manual if necessary. In most cases, the flat surface of the ferrule should touch the flat surface in the fitting or column nut; the tapered region should touch a tapered region in the fitting or nut.

Any time a ferrule is placed on a column, 2-3 cm from the end of the column must be cut off. There are no exceptions! Upon sliding the ferrule on the column, very small fragments are shaved from the inner surface of the ferrule by the sharp edge of the column end. These ferrule particles can enter the column and severely interfere with the chromatography. Peak tailing and adsorption are the most common symptoms of ferrule particles in the column. Sometimes, the particles migrate long distances into the column and permanently affect its performance. This is a greater problem with larger diameter columns.

The insertion distance of the column into the injector is an important measurement. The GC manual provides this measurement. Installing the column at the incorrect depth may lead to efficiency losses, changes in injector discrimination or peak shape problems. In some cases, an improvement may occur. The distance is usually stated from the end of column to a reference point such as the top of the ferrule or the bottom of the column nut. It is very easy to move the column from this distance during installation. Section 9.9.5 describes some methods to ensure the proper installation depth.

The carrier gas can be on or off during column installation. If it is on, this is insurance that the carrier gas is on during the later stages of installation where carrier gas flow is essential. Also, if the carrier gas is on, the injector fittings can be checked for leaks after the column is installed.

It is not necessary to cool the injector to install a column. As long as hot fittings can be tolerated, it may be best to leave the injector at a higher temperature. Fittings have tendency to leak when subjected to large changes in temperature. If a column is installed while the injector is cool, it is recommended to slightly tighten the column nut after the injector has reached the set temperature. The use of liquid leak detection fluids require the injector fittings to be below 100°C.

9.2.5 Turning On and Verifying the Carrier Gas Flow

Heating of a column without carrier gas flow results in serious and usually permanent damage of the column. Before the oven is heated after column installation, it is necessary to verify that there is a reasonable flow of carrier gas through the column. If there is a reasonable pressure on the gauge, it is often assumed that there is carrier gas flow in the column. This is not a completely reliable method. Flow controller or pressure regulator failures, or blockage in the gas lines or column can all create a pressure reading on the gauge; however, there is little to no carrier gas flow through the column. Damage to the column occurs upon subsequent heating. Worse is the damage to any replacement columns installed in the same GC with the same problem.

The best method to verify column flow is very quick and simple. Place the end of the column in a small vial containing a light solvent (e.g., hexane, methanol). If there is carrier gas flow through the column, a steady stream of bubbles is visible. An absence or a very slow stream of bubbles indicates a problem.   Check the GC system and gas supply, and make any adjustments or repairs before proceeding with column installation.

9.2.6   Installation in the Detector

Column installation in the detector follows all of the same guidelines and precautions as for installation in the injector (Section 9.2.4).   The proper alignment of the nut and ferrule, a good column cut, the proper insertion distance and leak free connection are all important.   Proper column installation in the detector is not as critical as for the injector, but great care should be taken anyway.

One of the carryovers from packed GC column usage is leaving the detector end of the column disconnected until the column is conditioned. While this is not necessary for capillary columns, it is still practiced.   This technique is discussed in Section 9.2.8.4

9.2.7   Verifying Column Installation and Detector Operation

A leak allowing oxygen (air) into the column or no carrier gas flow while the column is at higher temperatures (~50°C or higher) can rapidly damage a column.   After the column is installed in the GC, the next step is conditioning the column at a high temperature. Before heating the column, it is best to verify the existence of carrier gas flow (even though it was done previously). The easiest and most secure method is to inject a non-retained compound into the GC while the oven is at 40-50°C. A pressure on the column head pressure gauge is not always reliable. Do not heat the column above 50°C before carrier gas flow is verified. Before injecting the non-retained compound, the detector needs to be on and the column head pressure set at an appropriate value. Table 9-1 lists the approximate head pressure ranges for some common column dimensions and carrier gases so that the linear velocity is set close to the recommended values. The head pressure (linear velocity) does not need to be exactly adjusted and set at this time; this is done later when setting the average linear velocity.

The compounds used to verify carrier gas flow are the same ones used to set the average linear velocity (Section 6.4.1). Table 9-2 lists some recommended compounds. One advantage of injecting a non-retained compound at this time is the additional information that is gained. The presence of a peak for the non-retained compound confirms that the detector is on and that the data system is connected; the shape of the peak helps to confirm proper column installation. Avoid large volume injections or high concentrations since column overload can be misinterpreted as poor peak shape. Using a split injection works the best. If a peak does not appear, there is no carrier flow, the head pressure is set too low, not enough of the non-retained compound was injected, the detector is not functioning properly, or the recorder is not on or properly connected. It is absolutely necessary that a peak be obtained before proceeding. The peak obtained must be extremely sharp and without any tailing. If a broad or tailing peak is obtained, the following items should be inspected or checked:

1. Injector liner (breakage)
2. Position of the column in the injector and detector
3. Leaks in the injector and detector fittings
4. Leaks in the septum
5. Quality of column ends (i.e., poor cuts)
6. Ferrule or septum debris in the column
7. Split ratio (too low)
8. Inadequate makeup gas flow

Table 9-1: Approximate Head Pressures for Column Conditioning

| Diameter (mm) | Length (m) | Head Pressure (psig) | | |
		Nitrogen	Helium	Hydrogen
0.18-0.20	10-12	3-5	8-12	8-10
	20-25	6-10	16-24	16-20
	40-50	12-20	32-48	32-40
0.25	15	2-4	6-9	6-8
	30	4-8	12-18	12-16
	60	8-12	24-36	24-32
0.32	15	1-2	4-6	4-5
	30	2-4	8-12	8-10
	60	4-8	16-24	16-20
0.53	15	0.5-1	1-2	1-2
	30	1-2	2-4	2-4
	60	2-4	4-8	4-8

Table 9-2:   Non-retained compounds

Detector	Compounds
FID	methane, butane[1]
ECD	methylene chloride[2,3], dichlorodifluoromethane[2,3]
NPD	acetonitrile[3,4]
TCD, MS	methane, butane, argon, air

1.   A disposable lighter is a good source of butane. Place the syringe   needle into the flame outlet of the lighter. Depress the small button   allowing the gas to escape. Pull up 1-2 µL of the gas and inject.

2.   Headspace or diluted in solvent.

Headspace: Fill an autosampler vial with about 10 drops of solvent. Tightly seal the vial with a cap. Shake the vial for several seconds. Pierce the septum with the syringe needle, but do not insert the needle into the liquid. Pull up 1-2 µL of the headspace above the liquid and inject.

Solvent dilution: Add 25-50 µL of the appropriate solvent to 10 mL hexane or iso-octane and thoroughly mix. Inject 1-2 µL. Further dilution may be necessary depending on the sensitivity of the detector.

3.   Use a column temperature of 40°C or greater.

4.   Use a column temperature of 90°C or greater.

9.2.8   Column Conditioning
9.2.8.1   What is Column Conditioning?

All columns have a small amount of volatile contaminants originating from their handling, installation and storage. Column conditioning involves heating the column to a high temperature to rapidly remove these contaminants. Conditioning is necessary any time a column is installed even if it has been previously conditioned. Failure to condition a column results in baseline problems. The baseline severely rises as the column is heated and numerous peaks, humps and blobs may be visible.   While these contaminants do not foul or harm a detector, they are interferences in the chromatogram.

9.2.8.2 Conditioning Temperatures

There are two approaches concerning the best temperature to condition a column. One is at the isothermal temperature limit of the column or 20-25°C above the highest oven temperature that will be used (without exceeding the upper temperature limit). Using the upper temperature limit conditions the column faster, but it may shorten column lifetime by exposing the column to unnecessarily high temperatures. If the highest operating temperature is more than 50°C lower than the column's upper temperature limit, conditioning at 20-25°C above that operating temperature may be better. The drawback to conditioning at lower temperatures occurs if a higher column temperature needs to be used at a later date. The column usually has to be re-conditioned at a higher temperature to accommodate the new temperature conditions. Failure to re-condition the column may result in a rising and erratic baseline when the column temperature exceeds the previous conditioning temperature.

The rate at which a column is heated has no impact on column conditioning. Rapidly heating a column does not damage it. Thus slowly heating the column to the conditioning temperature is not necessary. A temperature program is not needed nor are repeated cycles of heating and cooling. The GC oven can be directly set at the conditioning temperature and the column heated at whatever rate the oven reaches this temperature. There is no evidence that temperature programming or cycling between temperatures improves conditioning or increases column lifetime or performance.

9.2.8.3 Conditioning the Column While Connected to the Detector

Occasionally there is a fear that the contaminants eluting from a column during conditioning will contaminate or foul the detector. While this is a valid concern with packed GC columns, it is not for capillary columns. The amount of stationary phase and contaminants in capillary columns is quite small. Also, most column manufacturers precondition the columns for prolonged periods as part of the production process. There are several advantages to conditioning a column while it is attached to a functioning detector. Conditioning a column while it is disconnected from the detector is not a problem; however, some information is lost and a short conditioning step is still needed. Use of this conditioning technique is further described in Section 9.2.8.4.

It is best to monitor the progress of column conditioning so that problems can be immediately noticed before column damage can occur. The detector should be already functioning due to the carrier gas flow verification step (Section 9.2.7). Set the recorder or data system with a 120 minute run (stop) time and the attenuation or sensitivity setting used for the analyses. Input the conditioning temperature into the GC and start the recording device. The baseline should start to sharply rise after several minutes and continue to rise beyond the time when the GC oven reaches the set temperature. After 5-15 minutes at the conditioning temperature, the baseline should start to rapidly drop. After 30-90 minutes at the conditioning temperature, a steady baseline or detector signal should be obtained. As soon as the baseline is stable for more than several minutes, conditioning should be stopped. Usually the entire plot looks like a large blob with some peaks riding on top (Figure 9-1). Further conditioning reduces the baseline by only a very small amount and usually is a waste of time. More polar stationary phases, thick film columns and greater sensitivity settings usually require longer conditioning times. For very low level analyses (sub ppb), 4-8 hours of column conditioning may be required.

Figure 9-1:   Baseline During Column Conditioning

If the baseline is unstable or excessively high after 60-90 minutes, column conditioning should be stopped. Most likely there is a leak in the GC or a contamination problem. If there is a leak, the baseline remains at a high level since damage to the stationary phase is occurring. The high baseline is due to the elution of the degradation products of the stationary phase.

Additional heating only further damages the column, thus conditioning should be stopped. The most common location of leaks is around the injector. Check the septum, column nut and all of the injector fittings for leaks. Be sure that the recording device is not set at one of its most sensitive levels. Normal but small deviations in the detector signal appear to be very large at these very sensitive settings.

Contamination is primarily visible as an erratic baseline and/or the presence of various size and shape peaks. The GC is usually not suitable for use while contaminated, thus conditioning should be stopped so that the GC can be cleaned. A dirty injector or gas lines are the most likely source of any contamination. In some cases, installation of a new column into a contaminated injector results in a contaminated column. If this contaminated column is transferred to a different, contaminant free GC for troubleshooting purposes, the same problem arises. The incorrect conclusion that the column is defective may be reached when in fact the column was contaminated by the first GC system. A new column should not exhibit high bleed. High quality columns do not self destruct upon shipping or storage, and are tested for bleed levels during the manufacturer's testing process.

Some detectors require several hours to stabilize especially if they have been off for more than 1-2 hours. The most common detectors that exhibit this tendency are the ECD, NPD and MS. Upon monitoring the baseline, it may appear that the baseline does not stabilize and continues to drift downward. This drift may be due to detector stabilization. The column maybe fully conditioned, but the detector requires additional time to stabilize. It is difficult to determine whether the drift is column or detector related. Experience with a particular detector usually helps in determining the source of the baseline drift.

9.2.8.4   Conditioning the Column While Disconnected from the Detector

Fears of detector fouling by contaminants eluting from the column during conditioning leads some chromatographers to leave the column disconnected from the detector. Fouling of the detector by the column is a very rare occurrence. Only PLOT columns and very thick film stationary phases have enough material to foul a detector.

Columns can be conditioned while they are disconnected from the detector without any problems. The baseline or background signal can not be monitored during conditioning. This makes it difficult to determine whether the column is fully conditioned. Also, ongoing damage to the column due to a leak at high temperatures can not be detected until significant damage has occurred.

If a column is conditioned while disconnected from the detector, 10-20 cm of the free end needs to be cut off before installation in the detector. A small amount of air enters the end of the column and damages the stationary phase in this region. After the column is installed in the detector, a small amount of additional conditioning is still necessary. The process of handling and installing the column slightly contaminates the end of the column. This requires the column to be held for 5-15 minutes at the conditioning temperature or until a stable baseline is obtained.

9.2.9 Setting the Carrier Gas Linear Velocity

The best time to accurately set the linear velocity is after the column is properly conditioned. The linear velocity can be set before conditioning; however, time is required for the column to stabilize (condition) at the temperature used to set the linear velocity. Since carrier gas flow has been previously verified, there is no need to set the linear velocity prior to conditioning.

The retention time of a non-retained compound is used to set the average linear velocity. A list of recommended compounds can be found in Table 9-2. The desired average linear velocity and column length is used in Equation 9-1 to determine the required retention time for the non-retained compound. The column head pressure is adjusted downward to increase the retention time and adjusted upward to decrease the retention time until the desired one is obtained. A retention time of ±0.05 minutes from the calculated value is sufficient in most cases. A table of recommended average linear velocities can be found in Section 6.6.

The average linear velocity is dependent on the column temperature. It is important to set the linear velocity at the same temperature for a given analysis. Using a higher temperature results in a slower velocity and a lower temperature results in a faster velocity. Changes in retention times and efficiency occur with the measurement error. Usually the initial temperature of a program is the most convenient. In practice, any column temperature can be used; however, a consistent temperature of measurement is the most important consideration.

After the linear velocity is set, the column is ready to use. There are several more steps that are not necessary for the best performance of the GC system; however, they will prove to be very useful upon the onset of future problems.

$$t_m = \frac{L}{60 \, \bar{u}}$$

t_m = retention time of a non-retained compound (min)
L = column length (cm)
$\bar{u}$ = desired average linear velocity (cm/sec)

Equation 9-1: Non-retained Peak Retention Time to Obtain a Specific Average Linear Velocity

9.2.10   Bleed Test

After the column is conditioned, a bleed test is one method to measure the amount of column bleed. As a column is used, the bleed level slowly increases. Eventually, it reaches a level that renders the column unsuitable for most uses. By having a bleed level measurement for the column when it was new, there is a reference point future comparisons.

There are two methods most commonly used to determine the amount of column bleed. The detector signal at the column's conditioning temperature is used as a measurement. The highest background signal should be obtained at the conditioning temperature or isothermal temperature limit, whichever is the highest. The signal can be obtained from the data system/recorder or detector output signal. It is better to compare this signal to the background signal at a lower column temperature (usually between 50-100°C). Contributions from sources other than the column are now factored out of the measurement. The difference between the two signals is used as the bleed level measurement.

Collecting a bleed profile is probably the best method to measure the bleed level. A bleed profile is generated by running a temperature program without an injection (i.e., blank run). The temperature program should follow these guidelines: initial temperature of 50-100°C, ramp rate of 10-20°C, final temperature equal to the isothermal upper temperature limit of the column, and a final hold time of 10 minutes. The appearance should be similar to that in Figure 9-2. The size of the rise in the baseline is the important feature. As long as the same scale is used to draw both chromatogram (before and current), the rise in a current bleed profile can be compared to the one generated when the column was installed.

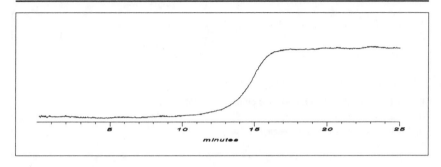

Figure 9-2:   Bleed Profile
Column: DB-1, 15 m x 0.32 mm I.D., 0.25 µm
Oven: 100-325°C at 15°/min, 325°C for 10 min

9.2.11   Injecting Column Test Sample

After the column has been conditioned and a bleed test run, some type of column test mixture should be injected and analyzed. There are two types of test samples. One is the same sample used by the column manufacturer to test a newly made column. The other is a representative or actual sample like the ones analyzed with the GC. Perhaps both can be used to test different aspects of the GC's performance.

Nearly every capillary column comes with a performance chromatogram generated using a specific test mixture. If the same conditions and sample are used, the chromatogram generated should be nearly identical to the one sent with the column. Large differences are indicative of a problem with the gas chromatographic system or perhaps the incorrect test conditions. When a good chromatogram is obtained, it should be stored for future reference. If a problem should arise in the future, the test mixture can be injected and the chromatogram compared to the one generated with the properly working system. The diagnosis of many problems can be accomplished with the proper interpretation of the results of the test mixture chromatogram. Each of the compounds in the test sample provides a specific bit of information about the column. This information is very useful when trying to troubleshooting column problems. Detailed information about test mixture compounds can be found in Chapter 10.

Instead of a manufacturer's test sample, a representative or actual sample can be used to test system performance. In some cases, this sample is a better test of the performance of the gas chromatographic system. Some analyses are less demanding of the column than the column test mixture. A column that appears to be bad with the column test mixture performs satisfactorily for the samples of interest. It is probably unreasonable to discard that column based on the results of the column test mixture when it adequately performs the desired analyses. Some analyses are more demanding of the column than the column test mixture. A column that appears to be good with the column test mixture performs less than satisfactorily for the samples of interest. While this is fairly uncommon, a different sample than the column test mixture is needed to adequately judge the performance of the column. Usually there are only a few compounds or peaks that place high demands on the column. The test sample should contain these compounds; the other sample compounds may not be necessary especially if they are not as demanding on the column.

One simple method to monitor system performance over time (especially if a large number of chromatographs or inexperienced users are involved) is to make an overhead transparency of the corresponding test chromatogram. The transparency can be placed over the test chromatogram generated subsequent to the initial instrument/column setup and performance verification. Acceptable performance variations can be noted on the transparency. Factors such as peak retention, shape and size, and baseline rise can all be easily noted and measured even the most inexperienced GC user. The need for system maintenance can be discovered before a major problem develops. Also, this helps to prevent the use of incorrect conditions or methods.

9.3   Column Ferrules

Graphite and graphite/vespel are used to make capillary column ferrules. Ferrules made of both materials are available for some GC's while others may be limited to only one. Essentially, ferrule material is primarily a personal preference. A few models of GC's require special column ferrules. In these cases, the exact ferrule is required since non standard fittings are often used for the injector and detector.

Graphite ferrules are very soft which makes them easy to seal. This makes it easier to get a leak free connection. The softness of graphite also creates a few problems. Upon overtightening, graphite ferrules sometimes generate small flakes of graphite. These small flakes can contaminate injectors, detectors, fittings and columns. The soft graphite flows into the threads and openings of the injector and detector fittings. Removal of this material is necessary before the next column is installed. Graphite ferrules are not recommended for use on the detector fitting for GC/MS and ECD systems. Small pieces of the graphite ferrule can contaminate detectors. GC/MS systems require hours of work to clean out particulate matter. The vacuum in a MS system also makes it easy to suck ferrule particles into the source. ECD cells are sealed, thus ferrule particles in the cell necessitate return of the detector to manufacturer and days of downtime. Overtightening a graphite ferrule makes it difficult to re-use the ferrule multiple times. A graphite ferrule can be used 10-15 times providing it is not deformed by repeated overtightening. Techniques to avoid overtightening fittings can be found in Section 9.4.

Graphite/vespel are much harder than graphite ferrules. This makes their properties the opposite of graphite ferrules. A leak free seal is more difficult to obtain with graphite/vespel ferrules since they are much harder. It is much harder to overtighten graphite/vespel ferrules. Graphite/vespel ferrules do not flake and are better suited for GC/MS and ECD systems. Graphite/vespel has the tendency to creep at higher temperatures. The ferrule changes shape upon heating. A leak free fitting at one temperature can leak at a different temperature. If the fitting temperature is changed by more the 25-30°C, it is recommended to check the fitting for any leakage after it has reached the new temperature. Upon overtightening especially at higher temperatures, graphite/vespel ferrules usually stick to fused silica tubing. Often the tubing has to be cut since the ferrule does not come off of the tubing.

Ferrule size is determined by the GC fittings and column diameter. The GC fittings governing the outer diameter and height of the ferrule. The most common outer diameter is 1/8" with a few 1/4" and metric sizes. This size does not refer to the outer diameter of the ferrule, but to the size of the fitting. The inner diameter is governed by the column diameter. There are three different ferrule inner diameters in common use. Table 9-3 lists the ferrule inner diameter need for the various column diameters.

Table 9-3:   Column Size vs. Ferrule Inner Diameters

Column I.D. (mm)	Ferrule I.D. (mm)
≤0.25	0.4
0.32	0.5
0.45	0.8
0.53	0.8

9.4   Tightening Fittings

Most fittings are overtightened. This is due to all of the cautions about avoiding leaks and the damage they cause. It is tempting to use a lot of force when tightening the column nut. While this does not cause any real damage, most ferrules can not be re-used if they are seriously overtightened. This can be quite expensive over the course of a year and a number of column installations. There are several guidelines that can be used to avoid excessive tightening of fittings.

Graphite ferrules are very easy to seal, thus little tightening of the fitting is needed. When using a new graphite ferrule, after tightening the column nut by hand, the column nut only needs to be turned about 1/4 of a turn using a wrench. Any more is using excessive force. A used graphite ferrule often seals in less than 1/4 of a wrench turn. It is easy to tell if the column nut is tight enough. If the column can not be pulled from the fitting, it will not leak. If the column moves during the pull test, make sure it is re-positioned in the proper location. Upon removal of the fitting, if the graphite ferrule is highly deformed or a portion has been forced through the opening in the fitting or column nut, the fitting was overtightened.

Graphite/vespel ferrules are hard, thus more difficult to seal than graphite. When using a new graphite/vespel ferrule, after tightening the column nut by hand, the column nut only needs to be turned about 1/2 of a turn using a wrench. Any more is using excessive force. A used graphite/vespel ferrule often seals in less than 1/2 of a wrench turn. If the column can not be pulled from the fitting, it will not leak. It is more difficult to overtighten a graphite/vespel ferrule. If the ferrules permanently stick to the fused silica tubing on a consistent basis, the fitting was probably overtightened.

9.5   Techniques of Measuring Column Insertion Distances

The proper position of the column in the injector and detector is important. The position of the column is usually controlled by maintaining a distance between the end of the column and a reference point on the column nut/ferrule assembly. It is not easy to maintain this distance while inserting the column into the fitting and tightening the column nut. As little as 2 mm of slippage can result in poor performance for some GC's. There are several easy techniques to ensure the column remains in the proper position throughout the installation process.

The most familiar procedure is to use liquid correction fluid ("White-Out") to mark the column. The column is positioned as noted in the instruction manual, the correction fluid is applied immediately below the column nut. The top of the mark aligns with the bottom of the column nut (Figure 9-3a). After the correction fluid is applied, make sure the ferrule does not drop or slip into wet fluid. Also, only a small amount of fluid is necessary to mark the column. The presence of excess material in or on the ferrule may prevent a leak free seal from forming between the column and the ferrule. If the ferrule slips into the wet fluid, remove the ferrule, cut off the marked portion of the column, and start the process again.

Another method of measuring column insertion distances utilizes a used disk-type GC septum. The column is inserted through the hole in the septum made by previous injections. The column nut and ferrule are then placed on the column. The septum is position so that the column nut and ferrule are in the proper location (Figure 9-3b). The septum fits snugly on the column and holds the column nut and ferrule in the proper place during installation. The septum stays on the column during use, but it does not affect the column in any manner.

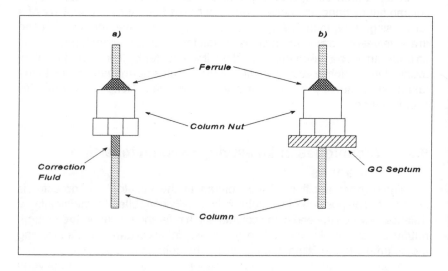

Figure 9-3:   Marking Column Insertion Distances

9.6   Leak Detection

Detecting leaks can be a difficult task. There are many places where leaks can develop with the injector fittings being the most common area. Liquid leak detectors are often used to detect leaks, but they do not work in many applications. Using an electronic leak detector is the most convenient, but they are very expensive. They work on hot fittings which is not possible with liquid leak detection fluids.

Liquid leak detection fluids are easy to use and inexpensive. The most common fluid is referred by the brand name of Snoop. When squirted onto a fitting, a leak becomes visible by the continuous formation of bubbles in the liquid. The absence of bubbles indicates that there is no leak present. A 50/50 mixture of iso-propanol and water can also be used as a leak detection fluid.

Liquid leak detection fluids do not work on hot fittings. Most detection fluids have an upper temperature limit of 90-100°C. Most injector and detector fittings are maintained at temperatures well above 100°C. To check the fitting for leaks, it has to be cooled below 100°C before the leak detector fluid can be used. Fittings often develop a leak upon heating especially if equipped with a graphite/vespel ferrule. If the fitting is leak checked at a lower temperature, tighten the fitting by about 1/16-1/8 of a turn after it has reached the higher temperature.

One method to determine whether an injector or the gas lines leading into it are leaking is to use a static leak check. This involves pressurizing the injector and gas lines and checking the column head pressure gauge for a loss of pressure. The details on this method can be found in Section 12.4.6.

10   COLUMN TEST MIXTURES

10.1   Column Performance Testing

One of the best means to evaluate the performance of a capillary column is by the injection of a column test mixture. Test mixtures contain compounds that have a range of chromatographic behaviors. Each compound in the test mixture usually provides a piece of information about the column. By evaluating the shape, height, width and retention of the test compound peaks, a substantial amount of information is obtained about the column and possibly the chromatographic system.

Practically every capillary column manufacturer individually tests each column using a test mixture and this test chromatogram is included with the column. The test sample used to evaluate the column can be very useful in the laboratory. It is useful in determining column performance and in many troubleshooting situations. It is important to save the test chromatogram and the accompanying information for future reference. This test chromatogram often provides additional information about the column such as the testing conditions, temperature limits, actual dimensions and serial numbers.

Depending on the laboratory and the analyses being performance, a test sample other than the column manufacturer's may be better. The manufacturer's test mixtures and conditions are designed to test the column and not the entire GC system. Test samples can be designed to evaluate the performance of the GC system. This is usually more desirable since it is the performance of the entire system that influences the final results. The manufacturer's test mixture and conditions may prove to be more valuable as a troubleshooting and column evaluation tool. Regardless of the situation or goals, some type of column or system test sample should be used.

10.2 Column Test Mixture Compounds
10.2.1 Hydrocarbons

Hydrocarbon peaks are the standard against which all other peaks are compared. Due to the lack of functional groups, hydrocarbons interact only with the column's stationary phase. Hydrocarbons peaks should be sharp and symmetrical without any tailing. Malformed hydrocarbon peaks are due to carrier gas flow problems in the injector or detector, solid debris in the column, very poor injection techniques, an extremely contaminated column or an extensively damaged stationary phase. Carrier gas flow problems include dead volumes or leaks in the injector, poor positioning of the column in the injector or detector, a broken or obstructed injector liner, insufficient make up gas flow, too low a split ratio or a poorly cut column end. If the amount of tailing decreases as peak retention increases, the problem is usually in the injector area.

Since hydrocarbons are not susceptible to activity problems, they are used to measure a number of column characteristics. The broadening of a hydrocarbon peak is solely due to the behavior of the hydrocarbon in the stationary phase, thus the hydrocarbon peak is used to measure the uniformity and thickness of the stationary phase. Hydrocarbons are used to measure the number of theoretical plates and coating efficiency since these values are affected by how well the stationary phase is deposited in the tubing. The partition ratio (k) of a hydrocarbon peak is used to measure retention, and thus the film thickness of the column. Straight chain (i.e., normal) hydrocarbons are used in the test mixtures with C_{10}-C_{16} being the most common ones.

10.2.2 Alcohols

Compounds with hydroxyl groups can interact with many materials through hydrogen bonding. Ideally the only interaction occurring inside the column is with the stationary phase. If this occurs, a symmetrical peak for the hydroxyl containing compound is obtained. In cases where an interaction between a hydroxyl containing compound and the tubing surface or contaminants in the column occurs, a tailing peak is obtained. A damaged stationary phase can also cause a tailing peak for a hydroxyl containing compound.

Straight chain alcohols are used in column test mixtures as sensitive indicators of column activity. Alcohols with chain lengths between C_8 and C_{12} are most often used. Diols are sometimes used in the mixtures as a more sensitive indicator of activity, especially for hydroxyl containing stationary phases. The most common diol used for this purpose is 2,3-butanediol.

Activity may also be caused by the presence of non-volatile or semi-volatile contaminants in the column or injector. The most common source of these materials are previous sample injections. Hydroxyl groups often interact with these residues, thus tailing peaks for alcohols and related compounds may not always be due to column activity. Contaminants causing activity problems can often be removed by solvent rinsing the column. If the alcohol peak tailing remains after properly solvent rinsing the column, either the tailing is caused by column activity or some of the contaminants are insoluble in the rinse solvents. In either case, the column is probably not suitable for use unless only inactive compounds are being analyzed.

10.2.3   Acids and Bases

Extremely acidic and basic compounds are among the most difficult to chromatograph since many of them are very active. Most column test mixtures contain both acidic and basic compounds to measure column activity and a column's pH behavior characteristics. Hydroxyl containing compounds are more sensitive to column activity problems than most acids and bases. In additional to column activity, acidic compound peaks tail if the column is too basic and the basic compound peak tails if the column is too acidic. In some cases, the acid or base peaks may be grossly distorted.

Adsorption is another problem with acidic and basic compounds. One of the differences between equivalent columns from different manufacturer's is the pH character. One manufacturer's column is better for acidic compounds because it is more acidic than another manufacturer's; therefore, the other manufacturer's column is better for basic compounds. To measure adsorption, the heights of the acid and base peaks are compared to the height of a hydrocarbon or FAME peak. Hydrocarbons and FAME's are not susceptible to adsorption, thus they are a good measure of the case where there is no adsorption. Height ratios below the established values indicate irreversible adsorption, thus a reduction in the peak height for the acid or base compound peak.

Dimethyl and chloro substituted phenols and anilines are the most common acid and base test mixture compounds. Dicyclohexylamine and 2-ethylhexanoic acid are often used when more sensitive basic and acidic compounds are desired. The peaks for these two compounds exhibit tailing with most columns.

10.2.4   FAME's

Fatty acid methyl esters (FAME's) are often included in column test mixtures. FAME's exhibit no appreciable activity which makes them suitable for many performance measurements. They are often used to calculate the number of theoretical plates, resolution and coating efficiency. Also, their retention indices are routinely used to determine whether the stationary phase selectivity and polarity are correct. Most column test mixture use the C_{10}-C_{14} methyl esters.

10.2.5   Other Compounds

Occasionally, there are some additional or different compounds in the column test mixture. Naphthalene or related compounds are found in some column test mixtures.   They are primarily used as a retention index compound. Additional compounds are sometimes added to the column test mixture for speciality columns. The compounds are often ones specified as test compounds in regulated methods or standardized testing protocols. These compounds are added to the column test mixture to ensure the proper performance (e.g., response factors, peak shape, retention) of the column for the specific method. Sometimes the special compounds are in a different sample, thus the columns are tested separately with this test mixture in addition to the standard column test mixture.

10.3   Column Testing Conditions

10.3.1   Injectors

Column manufacturer's are more interesting in testing column performance. The test conditions used for quality checking new columns are designed to minimize the amount of influence the GC system has on the final test chromatogram.   Column test mixtures are usually run using a split injector at a high split ratio. High split ratios minimize any injector related efficiency losses. Split ratios of 1:100 to 1:250 are used for most tests. Other injectors can be used if necessary, such as for low concentration samples or when exact analysis conditions have to be duplicated.

10.3.2   Detectors

Column manufacturer's use a flame ionization detector (FID) for their tests since it is very reliable and nearly universal in its response. This makes the test mixtures suitable for use with MS and TCD systems also. The use of other detectors poses a problem since many of the typical test compounds do not respond with a NPD, ECD or FPD. The presence of a responsive functional group on the test compound is required for selective detector use. Addition of halogens rarely changes the activity or important properties of compounds such as hydrocarbons or alcohols. The addition of nitrogen, phosphorus or sulfur often changes the activity of many compounds. This make its more difficult to find suitable test compounds for some of the more selective detectors.

Some specialty columns may be tested using a selective detector such as an electron capture detector (ECD) for a chlorinated pesticides column or a MS for low bleed columns. These tests are often run in addition to the standard column test mixture. In any case, it should be possible to duplicate these tests in the laboratory.

10.3.3   Column Temperature

Isothermal temperature conditions provide the best test of the column. A compound spends the same amount of time in every portion of the column. Every portion of the column has an equal opportunity to interact with each test compound. This is not the case with a temperature program. A compound spends more time in the front portion of the column than in the back half. The actual time spent in each portion depends on the temperature program (primarily the hold times and ramp rates). Problems in the front portion of the column have a greater influence on the chromatography than problems in the back portion. This is not equal opportunity testing of the column and it can create false impressions about its performance.

Temperature programming reduces the amount of visible peak tailing and creates peaks that are much narrower than isothermal conditions. Peak tailing is more evident using isothermal conditions. The primary problem with isothermal conditions is the need for test compounds that elute within a reasonable length of time. To obtain the best measure of the column, the isothermal test temperature should be greater than 90°C. If only the column is being evaluated, isothermal temperature conditions provide the best and most complete information about the column. If the purpose of the testing is to determine column or system performance for a specific analysis or set of conditions, a temperature program is satisfactory. A few column manufacturers use temperature programs while the rest use isothermal conditions in their column tests.

10.3.4   Test Sample Concentration

The concentration of a test sample is dependent on the injector, detector and intent of the test. Using high split ratios with a split injector requires the use of fairly concentrated test samples. Injectors that introduce a large amount of the sample into the column require lower concentration test samples. More sensitive detectors such as a NPD or ECD require the use of lower concentration test samples than less sensitive detectors such as a FID or TCD.

In general, lower concentration samples are more difficult to analyze. Peak shape problems, adsorption and baseline interferences are much greater problems when the GC system operates in a very sensitive mode. Many chromatographic problems are hidden when using concentrated samples. Peak tailing and small interferences are not as visible in the chromatogram since high attenuation values are used for highly concentrated samples. Problems are more visible at lower attenuation values. If a test sample that represents actual samples or conditions is being used, the sample concentration should be close to the concentration of the lowest level, actual sample to be run. If the system is working properly for the lower level samples, it will perform well for the higher level samples. If the test samples are at a higher concentration level, the system check reveals no problems; however, problems are experienced during the analyses (when it is too late) for the lower level samples.

10.4   Grob Test

The Grob test mixture is familiar to many gas chromatographers. It was first introduced around 1979 as an universal and demanding test for capillary columns. It is still used by some column manufacturer's for their column performance tests. If differs in several ways from the column specific test mixtures and conditions used by many column manufacturers.

Due to their widely different retention on most columns, a temperature program is needed to analyze the compounds in the Grob mixture. The Grob mixture uses the same compounds and temperature program conditions for every column. While having an universal sample and conditions is more convenient than individual test samples for each stationary phase then if any of the compound co-elute on a particular stationary phase, the information provided by those compounds is lost. A different temperature program can be attempted to better resolve the co-eluting compounds, but usually only small improvements are obtained.

The specific compounds in the Grob mixture are listed in Table 10-1. The Grob mixture compounds provide the same type of information as the isothermal column test mixtures; however, there are few differences. The n-decylamine, 2-ethylhexanoic acid and 2,3-butanediol are more active than most of the compounds used in other column test mixtures. Some non-Grob test mixtures contain some or all of these very active compounds. Slight tailing is visible for these peaks for nearly every capillary column. Also, the heights of the alcohol peaks are compared to a hydrocarbon or FAME; this value is not measured in most isothermal column test mixtures. The value of the trennzahl (TZ) is calculated using the FAME's in the Grob mixture. The TZ value is very dependent on column temperature, thus comparisons must use exactly the same temperature conditions. The trennzahl number is rarely calculated for the isothermal column test mixtures. Since the Grob mixture requires the use of a temperature program, retention indices are normally not calculated. Besides, retention indices are not as accurate for temperature programs as for isothermal temperature conditions.

The use of a temperature program can create a false impression of column performance. Temperature programs generate narrower peaks than isothermal conditions. Sometimes this creates the impression than one column is more efficient than another when the only real difference lies in the temperature conditions. Also, temperature program can conceal small amounts of peak tailing. It is more difficult to detect peak tailing for narrow peaks. Using a high chart speed for the chromatogram print out is suggested to better observe the active compound peaks for tailing behavior.

Table 10-1: Grob Mixture Compounds

Acids:	*Hydroxyl Containing:*	*Hydrocarbons:*
2-Ethylhexanoic acid	1-Octanol	n-Decane
2,6-Dimethylphenol	2,3-Butanediol	n-Undecane
	1-Nonanal	
Bases:	*FAME's:*	
2,6-Dimethylaniline	Methyl decanoate	
Dicyclohexylamine	Methyl undecanoate	
	Methyl dodecanoate	

10.5   Own Test Mixture

The best indicator of column and system performance is a sample containing some or all of the target compounds in the actual samples to be analyzed. The most relevant information is obtained in this manner. For example, if only hydrocarbons are being analyzed, the fact that the alcohols in the column test tail rarely affects the results of the actual hydrocarbon analysis. A test mixture containing only the important hydrocarbons is all that is necessary. Even though the column is not performing in the best possible manner (i.e., tailing alcohol peak), it would be premature to dispose of the column in this case. For this type of situation, the column test mixture is too rigorous for the actual analyses.

Some compounds are notoriously difficult to chromatograph and perfect chromatography should not be expected. A column that appears to be perfect with a column test mixture may exhibit tailing or adsorption for an actual sample. Perfect peak shapes and no adsorption are very difficult to achieve for low molecular weight primary and secondary amines, carboxylic acids (especially dicarboxylic acids), alcoholic amines, and some diols and dienes. Attempting to achieve perfect chromatographic performance with these types of compounds is usually a futile exercise. If these types of compounds are being analyzed, a test sample containing these compounds provides the most realistic and accurate measure of system performance.

Column test mixtures usually only provide efficiency measurements. Efficiency is not always a good measure of the amount of resolution for some closely eluting peaks. Slight differences in carrier linear velocities or other conditions can affect the resolution but have a much smaller impact on the measured column efficiency. An actual test sample provides this type of important information. The most difficult to resolve compounds can be included in the test mixture to make sure that critical resolution requirements are met. Other compounds such as those very susceptible to adsorption or tailing can be also included in the test mixture. These types of compounds often exhibit problems before the compounds in the manufacturer's test mixture. Routine checks of the system with an easy test mixture may not detect problems with the more difficult to chromatograph compounds.

Each analysis and laboratory probably uses a different set of performance criteria. Using an actual test sample to measure performance allows each method or laboratory to have its own set of testing protocols. For example, some methods may require high resolution, another needs good peak shapes while another needs sufficient peak response. The areas of the greatest importance or concern can be selected for monitoring while other unimportant areas can be virtually ignored. The areas of interest have to be selected based on previous analysis knowledge and experience.

However, the areas must be realistically selected and reasonable performance levels have to be set. This is to ensure that problems do not go undetected while normal variations within a method or technique are allowed.

10.6   When to Test a Column

It is recommended to test a column after installation, whether the column is new or has been previously used. Not only does this test the GC system before use, it also provides a performance reference point in the event of future problems. Upon encountering a problem situation, the test mixture can be injected, analyzed and compared to the original chromatogram for clues to the source of the problem. This makes troubleshooting much easier, especially if a column manufacturer's test mixture and conditions are used.

Some laboratories inject a test sample at regular intervals to insure the continued and ongoing performance of their GC systems. This verifies the proper operation of the GC system and increases confidence in the results. Whether this occurs before each set of samples, on a daily basis, or on some other schedule, is a decision that can only be made by the chromatographer or perhaps mandated by a regulatory agency.

11 CAUSES AND PREVENTION OF COLUMN DAMAGE

11.1 Causes of Column Damage and Performance Degradation

There are numerous misconceptions concerning the factors that can damage or affect the lifetime of fused silica capillary columns. Column technology has improved dramatically, resulting in capillary columns of greater durability, stability and performance. Columns are not indestructible, however, they still require proper care to maintain their high levels of performance. Some types of damage are permanent while others may be of a temporary nature. There are three primary causes of column damage - breakage, thermal and oxidation. The most common cause of column performance degradation is contamination of the column with sample residues.

11.2 Column Breakage

11.2.1 Causes of Column Breakage

Fused silica columns do not need to be handled with extreme caution and surgeon-like care. Fused silica tubing is flexible which allows it to be wound onto cages and easily installed in gas chromatographs. Rough handling of the column can result in breakage usually at a future date and not at the time of handling.

If the polyimide coating is abraded or scratched due to rough handling or contact with a sharp edge, the column is weakened at that location. Vibrations from the oven fan, continuous heating and cooling of the column or future handling may result in breakage of the column. Contact with identification tags, fan shrouds, metal edges in the oven or careless handling are all causes of premature column breakage.

It is a rare occurrence for a column to break without some sort of physical abrasion. The process of column manufacturing subjects the fused silica tubing to a series of harsh environments such as high temperatures, acid baths and tremendous amounts of physical manipulations. If a piece of tubing has a weak point, it usually breaks during the manufacturing process.

11.2.2   Symptoms of Column Breakage

The chromatographic symptom of a broken column is the lack of peaks after an injection. While other problems can cause the absence of peaks, it requires only seconds to look at the column for breaks. Depending on the location of the break, a drop in the column head pressure may be evident. The closer the break is to the injector, the greater the drop in the head pressure. Obviously, there are other problems or situations that can cause a drop in the head pressure.

11.2.3   Prevention of Column Breakage

The column should be placed in the GC oven so that the tubing is not subjected to excessive bending. Bends approaching 180° or greater subject the tubing to tremendous stresses at the bend points. These areas are stressed and are more likely to break than other portions of the column. The avoidance of excessive bends becomes more critical as column diameter increases. Large diameter tubing can not tolerate sharp turns or bends as well as small diameter tubing. Columns are slightly stressed from being wound on cages and the amount of stress on a column is dependent on the diameter of the column cage. Columns wound on small cages are more likely to break.

Megabore (0.53 mm ID) columns should not be wound on cages below seven inches in diameter or premature breakage may occur. The small winding diameter places excessive stress on the wider bore tubing. If less than seven inch cages are necessary, extreme care must be taken to minimize any other possible sources of tubing stress.

It is important that any points of contact between the fused silica tubing and column identification tags, oven walls, column hangers, etc. are eliminated. The forced air currents inside the oven causes movement of the column and these contact points may abrade the column. This leads to breakage at these abrasion points since this has the same overall effect as scribing the column for cutting.

11.2.4 Recovery from Column Breakage

If a column breaks, it is not rendered completely useless. A loss of 3-4 loops of tubing from the column does not significantly reduce its resolution capabilities. As long as a small decrease in the retention times is acceptable, install the longer length of the column back in the GC. If the same retention times are desired or needed, use an union to join the two pieces. If the break occurs further into the column, use a union to repair the break. Column performance is not sacrificed if the union is properly installed. No more than two unions should be installed in a column to repair breaks.

11.3 Thermal Damage

11.3.1 Causes of Thermal Damage

Exceeding a column's upper temperature limit for prolonged periods results in damage to the stationary phase. Short term exposure or temperatures slightly above the upper limit do not cause significant damage. There is more damage at higher temperatures above the column's upper temperature limit and the longer the time of high temperature exposure. Thermal damage most frequently occurs when two columns with different upper temperature limits are in the same column oven. One column is used without noticing that the analysis temperatures exceed the upper temperature limit of the other column. The second column is damaged since it is maintained at excessively high temperatures for prolonged periods.

Thermal damage also occurs when the column is heated without carrier gas flow. This usually occurs after changing a column or gas cylinder. The lack of carrier gas at temperatures above 50°C can result in damage to the stationary phase of some columns. In general, the more polar the stationary phase, the lower the temperature where significant thermal damage occurs.

11.3.2 Symptoms of Thermal Damage

Usually the first symptom of a thermally damaged stationary phase is excessive column bleed. The baseline rises to a very high level near the upper temperature limit of the column. The amount of bleed increases with greater thermal damage to the column. Peak tailing for active compounds occurs shortly after or simultaneously with the increase in column bleed. If the thermal damage is severe, all peaks tail with severe shape distortion being common. In some cases, a loss of efficiency and resolution occurs upon thermal damage. This is usually due to the distortion of the peak shapes. In most cases, excessive column bleed and activity problems render the column before efficiency losses become too severe.

11.3.3   Prevention of Thermal Damage

Thermal damage is avoided by not exceeding the column's upper temperature limit. The easiest method to avoid overheating a column is to set the oven maximum temperature on the GC to 5°C above the upper limit of the column. GC ovens sometimes exceed the set temperature by 1-2°C during rapid heating. If the oven maximum is not set slightly above the final temperature, the GC may shut off if only slight overheating occurs. Finally, the upper temperature limit can be prominently displayed on the GC as a reminder. Regular checks of the carrier gas cylinders and regulators insures that there is a constant supply of carrier gas.

11.3.4   Recovery from Thermal Damage

Most of the damage caused by overheating a column is permanent. If the overheating is not too severe, very little damage is noticed. In many less severe cases, only a small increase in column bleed occurs along with a small reduction in column life. If a more severe thermal damage situation occurs, a bakeout may return some of the column's performance. Disconnect the detector end of the column and supply carrier gas to the column. Heat the column to its isothermal temperature limit and leave it at this temperature for 12-24 hours. After the bakeout, cut off 10-20 cm from the free end. Install the column as usual and evaluate. The column will not return to the pre-damage performance level; however, it often is still useable. If the column is not suitable for use, it should be discarded since nothing else can be done to repair the damage.

11.4   Oxygen Damage
11.4.1   Causes of Oxygen Damage

At elevated temperatures, oxygen damages a stationary phase. The introduction of air into the carrier gas through a leak is the usual source of oxygen. Higher oxygen levels and higher column temperatures accelerate oxygen damage. Lower oxygen levels and column temperatures slowly degrades the stationary phase while higher oxygen levels and column temperatures rapidly degrade the stationary phase. Polar phases, especially Carbowax phases, are particularly sensitive to oxygen damage. A small introduction of oxygen such as with air injections do not harm the stationary phase. It is constant exposure to oxygen at elevated temperatures that causes the real damage to the stationary phase. Leaks and low grades of carrier gases are the most common sources of a constant exposure to oxygen.

11.4.2   Symptoms of Oxygen Damage

The symptoms of an oxygen damaged stationary phase are the same as for thermal damage. Usually the first symptom is excessive column bleed. The baseline rises to a very high level near the upper temperature limit of the column. The amount of bleed increases with greater oxygen damage to the column.   Peak tailing for active compounds occurs shortly after or simultaneously with the increase in column bleed. If the oxygen damage is severe, all peaks tail, with severe shape distortion being common. In some cases, a loss of efficiency and resolution occurs upon oxygen damage. This is usually due to the distortion of the peak shapes. In most cases, excessive column bleed and activity problems render the column useless before efficiency losses become too severe.

11.4.3   Prevention of Oxygen Damage

Avoiding leaks is the best method to prevent oxygen damage. Frequent and thorough leak checks are recommended especially after injector maintenance and column installation.   High purity carrier gases are recommended since they contain very low levels of oxygen. The use of oxygen traps is often recommended especially when using lower grades of carrier gas. The cost of the traps and the increased lifetime of the columns have to be compared to determine whether oxygen traps are worth their expense and extra maintenance. The cost of the traps may be justified in the savings realized by reduced column replacement when using lower grades of carrier gases.

11.4.4   Recovery from Oxygen Damage

The damage to stationary phases caused by exposure to oxygen is permanent. A small amount of damage may not render the column useless; however, the damage is not reversible. Sometimes, the high bleed level caused by the oxygen damage slowly drops with column use, but this bleed decrease is usually small. In most cases, the column needs to be replaced.

11.5   Chemical Damage
11.5.1   Causes of Chemical Damage

There are relatively few compounds that actually damage a stationary phase. Non-volatile compounds contaminate a column and can negatively affect the chromatography, but they do not physically attack and destroy the stationary phase. Some solvent/compounds combinations may not be compatible with a stationary phase and poor peak shapes are obtained. These solvents can not be used in these cases because of the peak shape problem and not due to any damage to the stationary phase.

11.5.1.1   Bases

Inorganic or mineral bases are particular damaging to stationary phases (Table 11-1). Substantial stationary phase destruction results if these types of compounds are introduced into the column. Damage occurs even if the base is neutralized by the addition of an acid prior to injection. There is always enough of the free base present in a neutralized sample to cause appreciable damage to the stationary phase. A low number of injections does not cause a significant amount of damage. Large numbers of injections, high pH and high concentrations of the base compound speed up the onset of damage related problems.

Organic bases such as alkyl amines do not harm the stationary phase since they a possess sufficient volatility to elute from the column. The inorganic bases are not volatile, thus they deposit in the column and remain for prolonged, if not indefinite, periods of time. This long residence time along with the strength of inorganic bases distinguishes inorganic from the organic bases in terms of the detrimental effects on column lifetime.

11.5.1.2   Acids

Inorganic or mineral acids cause damage to stationary phases (Table 11-1). Substantial stationary phase destruction results if these types of compounds are introduced into the column. Damage occurs even if the acid is neutralized by the addition of a base prior to injection. There is always enough of the free acid present in a neutralized sample to cause appreciable damage to the stationary phase. A low number of injections does not cause a significant amount of damage. Large numbers of injections, low pH and high concentration of the acid compound speed up the onset of damage related problems.

Organic acids do not cause damage to the phase since they possess sufficient volatility to elute from the column. High concentrations (>10%) of organic acids may cause slight phase damage especially if low column temperatures are employed. PEG stationary phases may take on the respective pH characteristic of the acid or base in the sample. Subsequent use of the column for samples of the opposite pH character often results in poor peak shapes.

Perfluoro organic acids are particularly strenuous on stationary phases. They can damage a stationary phase when present in high concentrations. Their usual source is as the by-product of the use of perfluoro acid anhydride derivatization reagents. Removal or reaction of the excess acid by-product is recommend to prolong column life and maintain performance.

Table 11-1: Compounds Causing Stationary Phase Damage

Inorganic or Mineral Acids:	*Inorganic or Mineral Bases:*
Hydrochloric acid (HCl)*	Sodium hydroxide (NaOH)
Sulfuric acid (H_2SO_4)	Potassium hydroxide (KOH)
Phosphoric acid (H_3PO_4)	Ammonium hydroxide (NH_4OH)*
Nitric acid (HNO_3)	
Chromic acid (CrO_3)	*Perfluoro Acids:*
Hydrofluoric acid (HF)	Trifluoroacetic acid (TFAA)
Perchloric acid ($ClHO_4$)	Pentafluoropropanoic acid (PFPA)
	Heptafluorobutanoic acid (HFBA)

* Condition dependent

11.5.1.3 HCl and NH_4OH

Hydrochloric acid (HCl) and ammonium hydroxide (NH_4OH) can be safely used under certain conditions. Hydrogen chloride is a gas at room temperature; however, it is usually dissolved in water as hydrochloric acid. Ammonium hydroxide is ammonia in water. Ammonia and hydrochloric acid are gases, thus they immediately elutes from most columns. Since HCl and NH_3 have little to no residence time in the stationary phase, damage does not occur. If water is present in the sample, HCl and NH_3 dissolve in the water. The HCl and NH_3 elute with the water. At column temperatures above 100°C, the water, HCl and NH_3 rapidly elute from the column, thus no damage occurs to the stationary phase. At column temperature below 100°C, the water condenses in the column. This traps the HCl and NH_3 in a concentrated form in the column. The long residence time of HCl and NH_3 results in damage to the stationary phase. Temperature programs starting below 100°C, but passing 100°C fairly rapidly are not a problem since HCl and NH_3 remain in the column for only a short time period. Isothermal column temperatures below 100°C or temperature programs remaining below 100°C for a prolonged time present ample opportunity for HCl and NH_3 to damage the stationary phase.

11.5.1.4 Organic Solvents and Water

Bonded and cross-linked stationary phases are not damaged by water or organic solvents. Even in the splitless and on-column injection modes where solvent actually condenses in the column, there is no damage to the stationary phase. Poor chromatography may result with some solvent, compound and stationary phase combinations, but no damage to the stationary phase occurs. Non-bonded and cross-linked phases exhibit some minor damage to the front part of the column after prolonged use in the splitless or on-column injection modes. About 10-20 cm of the front portion of a non-bonded and cross-linked column has to be removed periodically to maintain satisfactory and consistent performance.   For these types of columns, the use of a guard column is strongly recommended.

11.5.2   Symptoms of Chemical Damage

The first symptom of stationary phase damage by mineral acids or bases is peak shape problems. The stationary phase is destroyed in the front portion of the column where sample introduction occurs. The sample band is distorted which leads to peak shape problems. Also, active sites are formed and tailing peaks for active compound is common. In some cases, smaller peaks occurs due to adsorption. Since the stationary phase degradation only occur over a small region of column, the increase in column bleed is small. The bleed usually increases as the damage become more widespread. Loss of efficiency occurs if the damage is severe. The efficiency loss is primarily due to the peak shape distortion and not just a broadening in the peak width.

If the column has become damaged by acids or bases, there is probably a large among of the acid or base compound in the injector. The acids and bases causing damage are non-volatile or only slightly volatile. This means that an appreciable amount remains in the injector. Liner deactivation is lost by the action of the acid or base on the glass. If damage to the column has occurred from an acid or base, the injector should be cleaned and the liner replaced.

11.5.3 Prevention of Chemical Damage

The best method to prevent damage by acids or bases is to avoid injecting them into the GC. Often this is not possible due to sample preparation requirements, unknown sample origins and difficulties involved in removing the problem compounds. Attempting to neutralize a high or low pH sample does not prevent column damage. The problem acid or base is still present in solution. Also, the free acid or base often forms upon injection into the GC. Sometimes the acid or base added to neutralize the sample is as bad or worse for the column as the original problem compound. Ideally, the acid or base can be removed from the sample using some type of sample preparation method. Ion-exchange SPE cartridges are ideal for this purpose.

If a problem acid or base is present, there are several methods to reduce the extent of possible column damage. A guard column provides some protection against the action of the acid or base. Instead of collecting in the front of the analytical column, the acid or base collects in the deactivated fused silica tubing of the guard column. The very short resident time of the sample compounds in the guard column greatly reduces the contact between the compounds and the acid or base. By keeping the acid or base away from the stationary phase, much of the damage is avoided. Eventually, the guard column needs to be replaced as it becomes saturated with the acid or base compound. Peak shape problems or adsorption are usually the first signs that the guard needs replacement.

If a sample has to be pH modified for stability or to keep the compound in the proper form, use an organic acid or base, if possible. Acetic acid works well to acidify samples. Triethylamine is a good alkaline modifier. Similar compounds can be used if preferred. Peaks for acetic acid, triethylamine and similar compounds may be observed with some columns or conditions.

11.5.4 Recovery from Chemical Damage

Since stationary phase damage caused by chemical means is localized in the front of the column, the best method to repair the problem is to remove 1/2-1 meter of the front of the column. This removes the area of damage and is especially beneficial since the front portion of the column has the greatest impact on the chromatography. Solvent rinsing the column is of little help. While it removes the materials causing the damage, any existing damage is not repaired by the solvent rinsing procedure. At least further column damage is prevented upon removal of the acid or base compound by the solvent rinse.

11.6   Contamination
11.6.1   Causes of Contamination

The most common source of chromatographic problems is column contamination. These contaminants are materials that do not elute from the column within a reasonable amount of time. They accumulate in the column and cause a series of different problems. Contamination problems mimic almost every other GC problem. This can make it difficult to positively diagnose the problem. Contaminants can be classified into two groups: non-volatile and semi-volatile. It is possible to have both types of contaminants present at the same time. The symptoms of each type of contamination are similar, but there are a few differences.

Non-volatile residues are any portion of the injected sample that does not elute from the column. They accumulate in the column and injector, and it is this build up of materials from numerous injections that leads to the loss of performance. Anywhere from several to hundreds of injections may be made before the build up of residues interferes with the chromatographic process. The non-volatile residues deposit in the front portion of the column since they are not volatile enough to be transported by the carrier gas very far down the column. The majority of compound residence time in the stationary phase occurs in the front section of the column; therefore, residue contamination can have a very large influence on the quality of the chromatogram. This layer of residue causes several problems. It may interfere with the proper and efficient partitioning of the compounds between the carrier gas and the stationary phase. The contaminants may also prevent a sample band from properly forming at the front of the column. Finally, the contaminating residues can interact with the sample. In all of these cases, peak shape problems occur. Non-volatile residues should not cause any baseline problems since these materials are not eluting from the column.

Semi-volatile residues are contaminating materials that eventually elute from the column; however, it may be hours or days before they exit the column and cause any problems. These semi-volatile residues elute from the column and contribute to baseline disturbances and the presence of ghost peaks. They also cause peak shape problems for the same reasons as for non-volatile contaminants. Semi-volatile residues can contaminate substantial lengths of a column since they are mobile and can travel down the column. Some semi-volatile residues may originate from the thermal breakdown of non-volatile residues present in the injector or column. The high temperatures of the injector and column, and the repetitive cycling to high column temperatures are conducive to the breakdown of non-volatile residues.

The most common source of column contamination is the injected samples. Even samples that have been subjected to rigorous extraction procedures contain traces of semi- and non-volatile residues. With enough injections of a clean sample, a build up of contaminating residues can occur. A colorless or clear sample does not mean that there are no harmful residues in the sample. The brown materials that sometimes discolor injector liners are non-volatile materials. Obviously, some samples have larger quantities of contaminating residues than others. The use of selective detectors can contribute to contamination problems. There may be a large amount of compounds in the sample, but only those eliciting a detector response show up as a peak in the chromatogram. This can create the false impression that the samples are cleaner or less complex than in reality. Rapid contamination of the column can occur. If some of these non-detected residues elute late in the chromatogram, the GC run may be terminated before all of these compounds elute from the column. This leaves the contaminants in the column, thus they can interfere with subsequent analyses.

There can be a large number of compounds that do not gas chromatograph. These substances are introduced into the GC and accumulate until they cause a problem. Every compound in a sample does not have to be represented by a peak in a GC chromatogram. Just because there are six peaks in the chromatogram does not mean there are only six compounds in the sample. It means that there are only six compounds that are suitable for GC analysis in the sample.

Some types of samples are infamous for contaminating GC systems. Biological fluids and tissues, soils, sludge and waste water all contain high levels of non-volatile residues and can rapidly foul a column. Some samples like extracted drinking water, distilled solvents and petroleum distillate products contain very low amounts of non-volatile residues. These samples rarely contaminate columns. Other sources of contaminants can be non-sample related. These include residues or contaminants in sample vials and caps, solvents, pipets, septa and ferrule particles, and other non-sample related sources. Contamination sources of this type can be difficult to find and remedy.

11.6.2   Symptoms of Contamination

Column contamination problems mimic about every common GC problem. The most frequent symptoms of contamination are baseline disturbances (noise, drift, humps, blobs and peaks), peak shape problems (tailing, split, broad, and misshapen), and adsorption. Retention time shifts, inconsistent peak sizes and non-linear response can also be caused by column contamination.

It is often difficult to distinguish between column contamination and other problems. The use of the column test mixture is often helpful. The alcohols in the test mixture often exhibit obvious peak tailing. Sometimes the acid or base compound also exhibit peak tailing. The later eluting alcohol peak (if present) should exhibit more tailing than the earlier one. The hydrocarbons, FAME's or similar compounds are not affected by contamination unless it is fairly severe. If the alcohol peaks do not exhibit any tailing, contamination is probably not the problem.

If a temperature program is being used or can be used for testing, there are two quick contamination tests. If the presence of non-volatile contaminants is suspected, cut off 1/2-1 meter from the front of the column. Since most of the contaminants are located in this region, the most problematic area is removed. If the problem is substantially reduced, contamination is the probable cause of the problem. In some cases, removing 1/2-1 meter of the column completely eliminates the problem. Another test is to reverse the direction of the column. The contaminants are now located at the end of the column which is the least influential portion. If some improvement is noted, column contamination is a possibility. Care should be taken when using this technique. If there are significant amounts of semi-volatile contaminants at the front of the column, the reversal moves them very close to the detector. They may elute from the column and cause baseline disturbances which may complicate the evaluation results.

There is an easy method to determine whether an appreciable amount of non-volatile residues is present in a sample. Deposit about 10 µL of the sample on a microscope slide or small watchglass. Place the slide or watchglass on the injector port. The glass becomes heated to the approximate temperature that the sample experiences inside of the injector port. The solvent and any volatile sample components evaporate and the non-volatile portion is left on the glass surface. This residue is representative of the materials that may deposit and accumulate in the injector and front section of the column. If a large amount of residue is present, some type of sample clean-up should be undertaken to remove these residue before injecting the sample into the gas chromatograph.

11.6.3   Prevention of Contamination

The best defense against column contamination is to avoid having the contaminants in the sample. Good sample preparation techniques remove most of the problem substances. Even the best procedures and techniques do not remove every trace of each problem substance, thus contamination may still occur regardless of the prevention efforts. Some samples require extensive preparation or there are situations when sample preparation is not possible or reasonable. If proper sample preparation requires substantial labor, high cost, or unavailable equipment or expertise, tolerating column contamination may be the only choice. There are cases where it makes sense (economic and effort-wise) to frequently replace or clean columns instead of properly preparing the samples.

If there are unavoidable contaminants in the sample, the use of a guard column (Section 11.8) or packed injector liner (Section 11.9) may help. Both require periodic maintenance and can create other problems if not properly installed, used and maintained.

11.6.4   Recovery from Contamination

Cutting 1/4-1 meter from the front of the column often reduces or eliminates contamination symptoms, especially if the contaminants are of the non-volatile variety. If the non-volatile contaminants extend more the 2-3 meters into the column, it may be best to cut off 1/2-1 meter and then to reverse the column (only if a temperature program is being used). This can only be done once since reversing the column a second time only returns it to the original direction.

Solvent rinsing is the best method to remove contaminants from the column. Solvent rinsing involves flushing the interior of the column with a series of solvents; it is not injecting large volumes of solvent while the column is still installed in the gas chromatograph. It requires several hours of time; however, many columns return to their original performance levels. Also, it can be done numerous times without any harm or change to the column. Not all columns can be solvent rinsed, thus this needs to be determined before attempting any rinsing. If solvent rinsing columns has been repeatedly successful, it is well worth the time and effort. Two columns of the same description can be used for an analysis. As one becomes contaminated, it is replaced with the other one. The contaminated column is then solvent rinsed and used to replace the second column upon its contamination. The two can be used and rinsed many times until they fail from old age and long term use.

After solvent rinsing, it might be necessary to discard a column if it does not return to a suitable performance level. Before doing that, it may be possible to salvage half of the column. Much of the permanently contaminated or damaged area is in the front of the column. Cut the column in half and evaluate the back portion. About half of the time, the back half of the column performs very well. It is the front half that is responsible for the unacceptable performance. The short length may be satisfactory for the original analysis since excessively long columns are often used. If the half column is too short, the column may prove to be suitable for a different analysis. Perhaps, two of the shorter lengths can be joined with a union. This joined column may be a good backup in cases of emergency, a methods development column, or a column used to run very dirty or potential column damaging samples.

Baking out (keeping the column at a very high temperature) a contaminated column for 8 or more hours is often used to remove the contaminants. This is a common practice with packed GC columns. Capillary columns should not be baked out for any more than about 2 hours. Exposing the capillary column to continuous, elevated temperatures may cause some of the residues to polymerize in addition to shortening column lifetime. Then the only means to remove the polymerized materials is by removing the section of column containing the residues. In some cases, this means that up to five meters of column has to be discarded before chromatographic performance is be restored. In some cases, the column is no longer useable after a prolonged bake out.

11.7   Solvent Rinsing Columns

With use, capillary columns eventually become contaminated with residues. Often these residues can only be removed by solvent rinsing the column.   Solvents are forced through the column to wash out the contaminating materials. Only bonded and cross-linked stationary phases can be safely solvent rinsed. Do not rinse a non-bonded stationary phase or damage to the column will occur. Some column can be rinsed, but there are some solvents that should be avoided. Check with the manufacturer of the column if there is any uncertainty about solvent rinsing the column.

11.7.1   Solvent Rinse Kits

There are several methods to solvent rinse a column. The easiest method is to use a column rinse kit. They are available from nearly every capillary column manufacturer or a kit can be made for single or temporary purposes. The cost of the pre-made kits is low and they are very easy to use, thus it is simpler to buy one. The investment is easily recovered if only one capillary column is saved from unnecessary disposal.

Figure 11-1a shows the simple design and concept of a typical rinse kit. The detector end of the column is placed in the rinse kit vial through the straight arm of the tee-fitting. The column end should be located only 2-3 mm above the bottom of the vial. Some type of seal around the column is needed to prevent solvent or gas from escaping and to hold the column in place. Solvent is added to the vial and an inert gas is supplied via the other arm of the tee-fitting. Nitrogen, helium or argon can be used; air is suitable if other gases are not available. A pressure of 5-20 psi (30-140 kPa) is recommended. Longer and smaller diameter columns require high pressures. It is not advisable to exceed 20 psi since most rinse kits have some glass components. The pressure generated in the sealed vial forces the solvent to pass through the column. A beaker or similar container is used to collect the solvent as it leaves the column.

A home-made rinse kit can be constructed with readily available laboratory supplies (Figure 11-1b). All that is needed is a 10-20 mL screw cap vial, the corresponding open hole cap and septa, and about a 1 meter piece of fused silica tubing or clean, discarded column. Punch two small holes near the center of the vial septum. Fill the vial with the appropriate solvent and securely screw the septum and cap onto the vial. In the same manner as a regular column, install the piece of fused silica tubing or old column in the GC's injector. Insert the free end of the tubing through the septum so that it protrudes several millimeters into the vial, but not into the solvent. Insert the detector end of the column through the other hole in the vial septum. The end of the column should be 2-3 mm above the bottom of the vial. Using the flow or pressure controller of the GC, supply about 10 psi of pressure to the vial and adjust to maintain the proper solvent flow rate. This home-made kit works better for larger diameter and short columns since lower pressures are required. If the pressurizing gas tends to leak from the holes in the septum, either use smaller holes or use two septa in the cap.

A HPLC pump can be used to push solvent through the column. The column can be connected to the HPLC tubing using a metal union. A standard, graphite/vespel capillary column ferrule works in most metal unions. The desired solvents can be pumped through the column using the flow rate control of the HPLC. Keep the pressure below 50 psi for 0.53 mm I.D. columns and 100 psi for smaller diameter columns.

Another rinsing technique is to use a vacuum source to pull solvent from a reservoir through the column. Care must be taken not to pull solvents in the vacuum source. A trap or reservoir is needed to collect the rinse solvents.

Figure 11-1: Solvent Rinse Kits

solvents. Low boiling point solvents such as methylene chloride should be avoided when using a vacuum source. The generated vacuum reduces the boiling points of the rinse solvents and large amounts of solvent vapors can be produced. These vapors are more difficult to trap and can travel directly to the vacuum source. This is often damaging to vacuum pumps or contaminates the flowing water source used in a water aspirator.

11.7.2   Solvent Selection, Volumes and Flow Rates

The structure or identity of the contaminants in a column is usually unknown. For this reason, there are no exact guidelines to which rinse solvents to use for a particular situation. More than one solvent is often needed to remove the contaminating materials. A good general purpose rinse solvent selection uses methanol, methylene chloride and hexane.

These solvents have proven to work in most situations. Other solvents can be substituted; however, the new solvent should be similar in polarity to the one being eliminated.

There is a set of general recommendations concerning the rinse solvents. Using a range of solvent polarities is more successful. Try to use a polar, an intermediate polarity and a nonpolar solvent as part of the rinse scheme. If the previously injected samples were at a polarity extreme, the opposite polarity solvent can be eliminated. For example, methanol can be eliminated if petroleum samples (i.e., very non-polar) were analyzed. The best column rinsing occurs when each successive solvent is miscible with the previous one. For example, hexane should not be used immediately after methanol since these two solvents are immiscible. Another solvent such as acetone or methylene chloride should be used between the methanol and hexane since they are miscible with both of the other solvents. Another good choice is the sample or injection solvent. The contaminants are reasonably soluble in this solvent; thus they will probably be thoroughly rinsed from the column. Avoid solvents with boiling points above 100°C especially as the last rinse solvent. Since rinsed columns are heated to remove the last traces of solvent, a longer time is needed to condition the column than when using lower boiling point solvents. As a final rinse solvent, avoid halogenated solvents such as methylene chloride and chloroform if the column is going to be used with an ECD; avoid nitrogen containing solvents such as acetonitrile and DMF if the column is going to be used with an NPD. Trace amounts of these solvents create a high response with the specific detector. Very long column conditioning and detector stabilization times are required in these situations. If aqueous based or extracted samples have been injected (biological extracts, soils, waste water, etc.), include water as one of the rinse solvents. It is usually best to use it as the first solvent. Water and alcohols (methanol, ethanol, iso-propanol) should not be used to rinse PEG stationary phases. A small amount of stationary phase loss occurs especially if the column has been in service for several months.

The rinse solvent volume is dependent on the diameter of the column. Table 11-2 lists the recommended solvent volumes. Larger volumes can be used without any problems; however, this wastes solvent and time. When using multiple solvents, the previous solvent does not have to completely vacate the column before the next solvent can be started through the column. The next solvent can be added to the rinse kit vial or reservoir any time after the previous solvent is gone.

The flow rate of the solvent through the column should not exceed 1 mL/minute. For most columns, less than 1 mL/minute is delivered even at the highest recommended rinse kit pressure of 20 psi. Megabore (0.53 mm

I.D.) columns are the only ones that can easily accommodate flow at greater than 1 mL/minute. Smaller diameter and longer columns require longer times to complete rinsing since lower solvent flow rates are delivered at the maximum rinse kit pressure of 20 psi. Columns with 0.25 mm internal diameters and 60 meters in length may require up to one hour for each solvent to pass through the column. Viscous or heavy solvents such as water or methylene chloride flow at slow rates. More time is needed for heavier solvents to pass through the column than for light solvents such as hexane.

Table 11-2: Recommended Rinse Solvent Volumes

Column Diameter (mm)	Solvent Volume (mL)*
0.18-0.20	3-5
0.25	5-7
0.32	7-10
0.53	10-15

* Use the lower end of the range for 10-15 meter columns and the higher end in the range for 50-60 meter columns.

11.7.3   Conditioning the Column After Solvent Rinsing

After the column has been rinsed, it must be purged of any remaining solvent. Allow the solvent rinse kit pressurizing gas to flow through the column for about 10 minutes after the last solvent has left the column. Detach the column from the kit and install the column as usual. Do not heat the oven or condition the column at this time. Pass carrier gas through the column for 10-20 minutes.

After the carrier gas purge, verify that the injector is leak free. Condition the column using a temperature program. The initial temperature should be 40-50°C and the ramp rate at 2-3° per minute. Program up to the usual conditioning temperature for that column. Hold the final temperature until a stable baseline is obtained. Usually, obtaining a stable baseline requires 2-4 times longer after solvent rinsing a column than after a regular (i.e., non-rinsed) column installation.

11.7.4   Some Solvent Rinsing Considerations

There is no set limit to the number of times a single column can be solvent rinsed. In general, non-polar stationary phases can be rinsed more times than polar stationary phases. PEG stationary phases can tolerate the fewest number of solvent rinses. In most cases, the stationary phase degrades from normal use before the maximum number of rinses can be reached. PEG columns can experience a small amount of stationary phase loss upon solvent rinsing. This usually occurs when large volumes of polar solvents are used and for older columns.

If the same type of samples are routinely analyzed, the rate of solvent rinsing success should be considered. If three or more columns in succession are not saved by solvent rinsing, it is probably a waste of time and effort to continue rinsing columns in the future. If a moderate to large number of successes are experienced, future solvent rinsing of columns is usually worth the time and effort.

11.8   Guard Columns and Retention Gaps
11.8.1   Deactivated Fused Silica Tubing

A guard column or retention gap is a 1-5 meter piece of deactivated fused silica tubing attached to the front of the analytical column using some type of union. Shorter or longer pieces of tubing can be used, but 1-5 meters seems to be the most useful length. Guard column and retention gaps are installed for different reasons, but both provide the benefits of the other. Deactivated fused silica tubing is used for guard columns or retention gaps. It is the same tubing that is used to manufacture fused silica capillary columns, but it contains no stationary phase. The inner surface is deactivated to prevent any chromatographic interactions between the sample compounds and the tubing. Undeactivated fused silica tubing is available, but it is should not be used for retention gaps or guard columns. The diameter of the guard column or retention gap tubing is usually the same as that of the analytical column, but this is not mandatory.

11.8.2   Guard Columns

A guard column is used to provide a surface for non-volatile residues to deposit so that they do not accumulate in the analytical column. If the non-volatiles residues reach the analytical column, they usually coat the stationary phase and cause chromatographic problems such as peak tailing, splitting and adsorption. The vaporized portion of the injected sample passes through the guard column unretained since there is no stationary phase present. The residence time of the sample in the guard column is extremely short; therefore, a buildup of residues does not significantly affect the injected compounds. Since the non-volatile residues accumulate in the guard column, it can easily be discarded or a portion cut off when it becomes excessively contaminated. This eliminates the need to cut small lengths from the front of the analytical column as it becomes contaminated with residues.

Guard columns do not prevent semi-volatile residues from reaching the analytical column. The semi-volatile residues can easily move from the guard column into the analytical column since there is no stationary phase to retain the residues. The semi-volatile residues have sufficient volatility to be transported by the carrier gas into the analytical column. Non-volatile residues do not have sufficient volatility for this to occur.

The guard column has to be replaced when it becomes saturated with non-volatile residues. The frequency of replacement depends upon the types and number of samples injected. Usually a peak shape problem similar to those experienced without a guard column occurs. The guard column is saturated with residues; thus some of them may have overflowed into the analytical column. If this has occurred, a small amount of the front of the column needs to be cut off; 10-20 cm is sufficient in most cases. Careful monitoring of the number of samples injected before problems occur allows regular guard column replacement before complete guard column saturation has occurred.

11.8.3   Retention Gaps

A retention gap is used to improve peak shapes in specific situations. A retention gap is necessary for cool on-column injections especially when using secondary cooling since broad or tailing peaks are obtained without such a device. In some cases, narrower peaks are obtained with a retention gap when using splitless and Megabore direct injections. The most benefit occurs when there is a polarity mismatch between the sample solvent and the stationary phase.

If a non-bonded stationary phase is being used for splitless or on-column injections, a retention gap often protects the column. Sample solvent is collected and focussed at the front of the column. This disrupts a small amount of the stationary phase in this region since the phase is not bonded. After repeated injections, enough damage occurs to cause poor peak shapes. Solvent focussing occurs in the retention gap if one is installed. A retention gap also acts in the same way as a guard column; thus it may protect the column against non-volatile residue accumulation.

11.8.4   Unions

There are a variety of unions for attaching two pieces of fused silica tubing. Stainless steel or glass, press-fit unions are used most often. Each has certain advantages with the primary consideration being the ease of installation. If properly installed, unions do not have any detrimental effects upon the chromatographic results.

The low cost and relative ease of installation are the primary appeal of glass, press-fit unions. Their tendency to leak under certain situations is the main drawback. Proper installation is the best insurance against a leak. The fused silica tubing must have clean and evenly cut ends. The proper cutting tools and technique are required to obtain evenly cut ends (Section 9.2.2). Clean the end of the tubing by wiping with a laboratory wiper wetted with methanol or acetone. Clean the interior of the union by rinsing with methanol or acetone. Make sure the union is dry before using. The end of the tubing is inserted into the union until the tubing fits snugly into the tapered region of the union. The polyimide column coating compresses slightly and acts as the sealing force to hold the union in place and prevent leakage of carrier gas. Close inspection of the junction should reveal a continuous, dark ring around the tubing at the contact point with the union. If the ring is broken, the union and tubing do not have a proper seal and the connection will leak; re-installation of the union is necessary. If the union easily drops off of the tubing, it is not correctly installed. In some cases, the size of the tubing and the union are such that a good seal is not possible. If after several attempts to connect the union does not work, try another union. The normal size variances in unions and column outer diameters occasionally cause a specific union and piece of tubing to be incompatible.

Sometimes upon repeated heating to 200-250°C or above, a glass union and fused silica tubing permanently fuse together. Fusing does not always occur, thus it can not be relied upon to form a leak free seal. If a glass, press-fit union is used in a GC where the initial and final temperatures in a temperature program are far apart, or to connect a transfer line in a GC/MS system, leaks can develop upon use. A permanent and leak free connection can be made using polyimide sealing resin on the union, but the proper technique is essential. Clean the union with acetone. Lightly scribe the column 1-2 cm from the end, but do not break the column at this time. Rinse several centimeters of the outer surface of the tubing with acetone while holding the open end of the column upward. Using a laboratory wiper (such as a Kimwipe™), apply a thin layer of polyimide sealing resin around the column over the previously scribed portion. While still holding the column with the open end upward, break the column at the scribe mark. Inspect the cut end with a magnifier. If the cut is unsatisfactory (uneven or jagged) or resin has gotten into the column, repeat the cutting and resin application process. Upon a satisfactory cut, place a small drop of resin on one side of the tubing about 0.5 centimeters from the end of the column. Insert the tubing into the union and press firmly (but not so hard as to break or chip the column end). On the side where the drop of resin was applied, the gap between the tubing and the union should be filled with resin. If a gap is present, add just enough resin to fill the gap, but do not fill the entire gap around the column. Carefully install the assembled column into a GC and establish carrier gas flow. Carefully examine the union for any slippage or looseness. Heat the tubing for at least 30 minutes at 150°C to cure and set the resin. Longer times are not necessary, but there are no problems associated with the longer cure times. If the union needs to be removed, the tubing can be cut adjacent to the union.

Stainless steel unions can be used instead of glass, press-fit unions. A better seal is usually obtained when compared to glass unions without polyimide sealing resin. Stainless steel unions can be difficult to properly install. They require the use of very small ferrules and it is practically impossible to visually determine if there is a gap between the tubing ends. A gap contributes to peak broadening or tailing due to the excessive dead volumes especially for small diameter columns. The column has to be installed and tested before any problems can be detected. It is not unusual for several attempts to be needed before a good connection is obtained. One union design uses a tapered glass insert so that the column ends can be positioned with a minimum of dead volume. The high thermal mass of stainless steel unions may pose some problems when rapid and extreme heating or cooling of the column oven is involved. The union does not cool or heat at the same rate as the tubing. Peak shape distortions and retention time reproducibility problems may result. Some type of clamp or holder is usually required to hold the heavy union in place. A metal union that is not held in place may lead to column breakage or pull a portion of the column onto the bottom of the GC oven. A final consideration is that the union ferrules often cost more than a single glass, press-fit union. In some union designs, the ferrules are not easily re-used, thus replacement ferrules are always necessary.

11.9 Packed Injector Liners

If removing the majority of the harmful residues from the samples prior to injection is not possible or requires too much effort, using the injection liner as a trapping device may help to delay the onset of performance problems. A loosely packed plug of silanized glass wool inserted in the injection liner can act as a filter or adhesion site for some of the non-volatile residues and this reduces the amount reaching the column. Packed injector liners should be used with split injections and in high carrier gas flow Megabore direct injections; loss of efficiency often occurs with splitless injections.

The glass wool should be packed tight enough to filter out the residues without causing excessive restriction of the carrier gas flow. Distorted peak shapes or quantitation problems are signs of a too tightly packed plug of glass wool. Do not crush the glass wool into a tiny ball. Gently fold the glass wool fibers into a spherical shape and place the plug in the liner. Do not use regular lab grade glass wool. Injection liner glass wool must be silylated or irreversible adsorption and poor peak shapes occur. It is very difficult to correctly and completely silylate glass wool in the laboratory. It is recommended to purchase the silanized glass wool from a chromatography products supply company to ensure the proper material.

For extremely dirty samples, a different type of packed liner can be used with split injections. A small amount of silylated glass wool is placed in the liner followed by 2-3 mm of packed GC column packing material. A 1-3% loading of OV-1, OV-101 or a similar stationary phase is recommended. Another small plug of silylated glass wool is placed on top of the packing to hold it in place. The packed column packing traps the contaminants The packing has to be routinely changed since it becomes dirty with use. Make sure the injector temperature does not exceed the upper temperature limit of the packing material. It is recommended that a newly packed liner be heated in the injector with carrier gas flow before the column is installed. Small amounts of the stationary phase from the packed column material may bleed off and contaminate the capillary column. Several hours at the injector temperature is usually sufficient to condition the liner packing.

11.10 Gas Impurity Traps

One way to prolong column life is to minimize the amount of oxygen in the carrier gas. The use of high purity gases or the installation of an oxygen trap in the carrier gas line are the easiest methods. Moisture or water traps are also used in many systems since water can contribute to a higher background signal especially when using water sensitive detectors.

If a water trap is being used, it should placed before an oxygen trap. Oxygen is not removed by a water trap; however, water is retained on an oxygen trap. If the oxygen trap is placed first, its capacity can be prematurely reached by the presence of water in the gas. Water traps have very high capacities and are often inexpensively re-filled. Oxygen traps are much more costly than water traps. Oxygen traps have much shorter lifetimes, are not re-fillable, and require more frequent changing. The cost of a water trap is quickly returned by the savings in oxygen traps.

Installation of an indicating oxygen trap is often recommended if routine changing of the oxygen trap does not occur. The indicating oxygen trap is installed after the regular oxygen trap and is intended only as a warning that the regular oxygen trap has expired. An indicating oxygen trap has low capacity and is not intended to remove large quantities of oxygen. It quickly expires if a regular oxygen trap is not in place. The indicating trap can be very important since an expired oxygen trap is worse than no trap at all. An expired trap may leach or dump materials back into the carrier gas that often causes high background signals or contamination peaks.

Glass or metal bodied traps are preferred over plastic bodied traps. Some detectors such as ECD's or mass spectrometers are very sensitive to impurities that may be present in the plastic trap bodies. Some plastics may be slightly permeable to oxygen. Traps are usually used just for the carrier gas. A water and oxygen trap is often recommended for the auxiliary or makeup gas for ECD's. Most other detectors are not significantly affected by the presence of water or oxygen in their gases. There is one oxygen removal system that uses a catalyst type of an action. This eliminates the need to replace oxygen traps on a regular basis. The initial cost is much higher, but there is no further expense until the catalyst needs replacement in several years.

Traps should be placed in a vertical position. When placed horizontally, it is possible for some of the packing material to settle downward and form an open channel near the wall of the trap. Most of the gas goes down the channel instead of passing through the packing material. Water or oxygen is not removed when this happens. The traps should be placed as close to the GC as reasonably possible and they need to readily accessible for inspection and changing. If multiple GC's are on the same gas supply line, the traps can be placed before the point where all of the gas lines branch off to the individual systems. Minimize the number of gas line fittings past the traps. Each fitting potentially reduces the quality of the gas and counteracts the activity of the traps.

Hydrocarbon traps are available to remove trace levels of volatile organic compounds from the gas. These traps are packed with activated charcoal, thus they trap more than just hydrocarbons. Hydrocarbons traps are used where problems with organics from improperly cleaned gas lines, low quality gases, and compressor supplied gases are used. Hydrocarbons traps are not necessary unless contamination problems have been experienced.

Some GC's have small, internal traps, usually packed with activated charcoal or molecular sieves. These traps need to be replaced or rejuvenated periodically. The recommended procedures and internal trap locations can usually be found in the instruction manual for the gas chromatograph.

When using the higher quality grades of carrier gases, the use of gas impurity traps is debatable. The levels of oxygen and water are so low in the higher grades of carrier gas, the traps do not improve the quality of the gas. While the traps do last a long time, the cost, maintenance and leak concerns with the extra fittings may not be worth any of the possible gains. Traps are often not used when using the higher grades of carrier gas. The extra cost of the higher grade gas is often offset by the savings in not using traps. Also, the higher maintenance (inspection and changing of the traps) and leak possibility with traps is eliminated. Traps are necessary for the lower grades of gases. The traps reduce the levels of water and oxygen, thus improving column life and, in some cases, detector response. For situations where columns are frequently replaced and discarded, the use of traps may not be economically reasonable. No gain in column life is obtained and, as long as sensitivity is not affected by the absence of traps, installing gas traps is an unnecessary expense.

11.1 Column Storage

Columns should be stored such that scratching or abrasion of the fused silica tubing is avoided. The original column box is perfect for storage purposes. The column should be sealed by placing a GC septum over each end. Old, used septa can be used. Many fused silica capillary columns used to be flame sealed using a torch. It has been found that this is not necessary and sometimes did more harm than good. Columns do not need to be flushed with carrier gas before storage.

The shelf-life of columns is not completely known. Non-polar stationary phases have exhibited no performance degradation even after five years of storage. There have been isolated incidences of new columns with very polar stationary phases exhibiting slight performance changes after years of storage. The changes are small and often do not cause any problems.

11.12 Column Repair

Directly stated, damaged capillary columns can not be repaired. Columns can not returned to the manufacturer for repair since no column manufacturer accepts columns for repair. The cost and effort of repairing a column (if possible) is nearly the same as producing a new column. It is much better and easier to obtain a new column. Besides, many types of damage are not reversible and can not be fixed. There are some rejuvenation solutions available that claim to return capillary column performance. These solution rarely work and any improvement is very short lived.

12   TROUBLESHOOTING CAPILLARY GC SYSTEMS

12.1   Introduction

Any chromatographic problem can often be quickly resolved if a logical and efficient approach to the problem is maintained. Usually any superfluous amount of panic, fitting tightening, system alterations, random button pushing and knob turning only makes the actual problem more difficult to isolate and remedy. It is important not to jump to conclusions or make assumptions based on incomplete data. It is very easy to get fooled if some of the facts are not available or ignored. The best allies when troubleshooting are a good understanding of the basics of gas chromatography, capillary GC columns, the proper operation and characteristics of the particular model of gas chromatograph in question, the proper tools and common sense. Deficiencies in any of these areas make problem solving a more difficult undertaking.

One difficulty in troubleshooting is the multiple cause and effect nature of many GC problems. A single symptom can be caused by many different and non-related problems. Also, a single problem can have multiple symptoms some which are not evident in all cases. Finding the real problem requires the proper approach and evaluation of the situation.

Most problems are evident in the chromatogram; however, the chromatogram is the last link or operation in the chromatographic process. This means that the source of the problem can be anywhere or anything before the data system or recorder. In some cases, the recorder or data system is the source of the problem. Since many problems are only evident via the chromatogram, logical and simple procedures to find or eliminate possibilities are required to quickly remedy any problem situation.

12.2   Approaches to Solving GC Problems
12.2.1   Systematic Approach

One of the most important aspects of solving problems with capillary GC systems can be condensed into two words - Systematic Approach. It is often forgotten that the gas chromatograph and capillary column function as a complete system and not as two individual parts. In addition, the gas chromatograph can be viewed as constructed of a number of smaller systems or parts (injector, detector, etc.). Many of the small parts can be broken down even further. A problem or deficiency in any part of the GC including the column often results in some type of chromatographic difficulty. Using a logical and controlled troubleshooting procedure to quickly and accurately identify the source of the problem is the systematic approach. Maintaining the realization that the entire gas chromatographic system is a collection of different, but interrelated parts makes troubleshooting an easier and less complicated enterprise.

12.2.2   One Thing At a Time

There is a very strong and unconscious urge to change more than one parameter or variable at a time. Also, there is the urge to take everything apart before examining and evaluating the situation. Trying to examine and repair too many things at once is often less efficient and more time consuming than using a series of simple and conclusive tests. If numerous variables are simultaneously altered, the corrective action that repaired the problem is not known. While fixing the problem is important and the ultimate goal of troubleshooting, knowledge about the actual source and cause of the problem is better. Armed with this knowledge, preventative measures can be undertaken so that the problem does not re-occur in the future, and if it does occur, the proper actions can be immediately taken to repair the problem.

12.2.3   Moving from the General to the Specific

Do not try to solve a problem with a single, all-encompassing test, experiment or deduction. Determine the most general possible causes for a problem and then start to narrow the focus on specific causes. As an example, a problem with a loss of resolution occurs. A loss of resolution is caused by a decrease in peak separation and/or an increase in peak width; thus, the real problem is a separation and peak width issue. First, investigate whether a separation decrease or peak width increase (or both) has occurred since this is the general problem. Next, determine the possible causes of whichever problem was detected - this is focussing on the more specific problems. This approaches helps to focus attention on the causes of the real problem instead of the symptoms. Jumping to conclusions without finding the real problems often wastes time and effort since the wrong problems are investigated.

12.2.4   Checking the Obvious

Often the most obvious problems or solutions are overlooked because of their simplicity. Usually the simplest items are the easiest ones to check. Even if a setting or measurement was made while the GC was functioning properly, check them again after a problem occurs. Accidental and inadvertent changes occur without anyone's knowledge. Many of the areas that need to be checked seem trivial, but too many times it is deficiencies, malfunctions or incorrect parameters in these areas that are responsible for the problem. Some areas can be instantly eliminated using common sense. For example, if a chromatogram is being generated, it is very likely that the data system is connected properly and the detector is on. If no peaks are generated with an injection, both of these areas are possible causes of the problem.

The usual areas of first inspection and questions that need to be answered are:

1. Gas supply - correct pressures, correct gases
2. Gas flows - carrier gas linear velocity, split vent, septum purge, splitless purge, makeup and detector combustion/auxiliary
3. Temperatures - column, injector, detector and transfer line
4. Leaks - fittings, septum, gas lines
5. Is the detector turned on?
6. Using the intended injector and detector?
7. Sample - concentration, injection volume, solvent
8. Column - correct one, broken, installation
9. Syringe - plugged, different syringe
10. Data system - correct channel, settings, connections
11. Gas impurity traps - expiration, leaks, installation

12.2.5   Looking for Changes

Be conscious of any changes in any area related to the analysis. Some changes are very obvious such as a new gas cylinder or column. Other changes can be more difficult to detect or notice. For example, vial or cap supply, solvent brand or bottle, syringes, pipets, extraction columns, GC septa, the person making the injections have all been causes of changes in the chromatographic results. Some of the smallest and most trivial changes can affect the results. Changes occurring outside of the laboratory are often not noticed or reported, thus investigations of the onset of a problem can lead to areas outside of one's control.

12.2.6   Eliminating the Possibilities

The easiest way to get to the source of a problem is to reduce the number of possible causes. Instead of trying to find the source of the problem, try to eliminate the sources that can not be the cause. One of the remaining causes is probably the reason behind the problem. Simple experiments or tests are the best. Trying to eliminate too many choices with a single test often confuses the situation. Common sense usually dictates the plan of attack. As an example, some peaks are missing in the chromatogram. Areas such as the detector being on, the presence of carrier gas flow, correct data system connection can be eliminated since some peaks are observed. Problems in these areas would result in the complete absence of all peaks. If the retention times of the peaks are normal, problems in areas such as carrier gas linear velocity and column temperature can be eliminated. Changes in these areas would result in a change in the retention times. If the sample is injected into another suitable GC system, a large amount of information can be gained. If the same type of results are obtained, the sample and the syringe can be eliminated as possibilities. This has narrowed the possibilities down to the injector or column. Installation of a verified good column and its proper performance can eliminate the original column as the source of the problem. With a little thought and a few simple tests, many possible problem areas are eliminated. Another method to eliminate possibilities is to divide as much of the GC system into individual components and evaluate each component. By eliminating each component, the one causing the problem should be eventually discovered.

12.2.7   Divide and Conquer

By dividing the GC system into separate components, the performance of each component can be evaluated. If a component functions properly, it can usually be eliminated as the problem source. A divide and conquer approach works very well when there are multiple units comprising the GC system. As an example, peak shape problems occur with a GC equipped with a purge and trap sampler. Introduction of a known standard using the vaporization injector instead of the purge and trap sampler divides the system into two sections. If the problem goes away, the source of the problem is associated with the purge and trap sampler; if the problem remains, the problem is probably with the GC or column. By removing the purge and trap sampler from the system ("divide"), it is easy to determine whether it is responsible for the problem ("conquer"). Dividing and conquering is eliminating possibilities in a systematic manner.

There are two approaches to dividing a system. One is to split the system into two equal parts (as much as possible) so that the largest sections are examined. This is often not possible due to the design and operational requirements of many GC systems. The other approach is to start at the data system and work towards the injector and gas supply. After the data system is investigated, the detector is checked. This continues until the problem is found. This is often necessary in simple GC systems with very few options, accessories or operational possibilities.

12.2.8   Asking "If...Then..." Questions

A good foundation in basic GC theory and common sense are very powerful GC troubleshooting tools. Before attacking a problem, asking and answering some questions can often help focus the search for the problem. The best types of questions (after the obvious ones in Section 12.2.4) are the "if...then..." variety. For example, *if* column activity is the cause of the peak tailing, *then* the alcohol peaks in the column test mix should tail but not the hydrocarbon peaks. If the alcohols or similar compound peaks do not tail, column activity is probably not the cause of the peak tailing. If all of the peaks tail, then column contamination is a possibility. Another test needs to be run or a new set of "if... then..." questions need to be asked. "If...then..." questions are very powerful and can often severely narrow the possible problem causes. Very effective troubleshooting tests can often be built around "if...then..." types of questions.

12.2.9   Looking for Trends, Patterns and Common Characteristics

Sometimes the presence of a common trait or characteristic provides useful troubleshooting information. As an example, there is a noticeable pattern of decreasing peak tailing with increasing peak retention. In general, the earlier the peak retention time, the more volatile the respective compound. Thus, the trend in this example is the more volatile the compound, the worse the peak tailing. This points to a problem in the injector or carrier flow behavior since highly volatile compounds are more susceptible to expansion, backflash and vaporization problems. Activity is probably not the problem since peak tailing becomes worse with increased peak retention. Also, only the active compounds would exhibit peak tailing.

A set of patterns and trends can often help to narrow the list of possible problems sources. Characteristics such as compound structure, volatilities and solubilities can provide clues about the cause of the problem. Some trends or patterns are very easy to detect while others are nearly impossible since so many variables can be involved. Simple items such as sample color, brand of vial cap, time of day, number of previous sample injections are easy to find after a little investigation. Others such as sample exposure to light, position in a sample evaporator, ambient air humidity and contamination by solvent extractions performed in an adjacent laboratory can be very difficult to find and verify.

12.3   Tools and Useful Items

Maintenance and troubleshooting are easier when the right tools are readily available. Some tools are indispensable for successful trouble-shooting and also for maintaining capillary GC systems in the best working order. Every gas chromatograph health care package and first aid kit should contain:

1. *Flow meter.* A digital or manual model is suitable. The volume range of the meter should be compatible with the values being measured. For example, using a 10 mL flow meter to measure 100 mL/min may result in irreproducible and inaccurate flow values.

2. *Check-out column:* This is a column of known, verified and documented performance and quality. It is used to check the performance of the GC or to eliminate the original column as the problem source (Section 12.4.3).

3. *New syringe*: A clean syringe exclusively reserved for troubleshooting purposes. This helps to determine whether the original syringe is delivering the correct volume or is contaminated.

4. *Non-retained compound*: A non-retained compound is used to measure and verify the linear velocity of the carrier gas.

5. *New septa, ferrules and injection liners*: These are used to replace parts that eventually become defective, worn out or dirty with use.

6. *Column test mixture or reference sample*: Test mixtures are used to diagnose select system problems. They are also useful to compare current system performance with previous, satisfactory performance. With the correct interpretation, test mixtures are very powerful diagnostic tools.

7. *Leak detector*: An electronic detector is the best (but most expensive) way to find leaks. Liquid leak detection fluids are satisfactory for detecting leaks in non-heated areas.

8. *Log or record book*: Log books are invaluable for determining the past history of the system (e.g., number and types of samples injected, septa changes, occurrence of problems, maintenance schedules). A log book makes it easier to spot trends or to determine the frequency of required periodic maintenance. The log book is also a good place to keep performance specifications or requirements.

9. *Instrument manuals*: These are not a last resort. Manuals are a very useful resource for determining system performance specifications, proper operating guidelines and instructions on the disassembly of hardware like injectors and detectors. Instrument manuals are a good source of troubleshooting information specific to a particular GC.

12.4   Troubleshooting Tests
12.4.1   Jumper Tube Test
The jumper tube test is a good general purpose test to find or eliminate possible problem sources. It is especially helpful in tracking down contamination, baseline drift or noise problems of unknown origin. There are three steps to a jumper tube test. First, remove the column and cap off the

entrance to the detector. Turn on the detector as usual with all of the normal detector gas flows and temperatures. Activate the detector by igniting the flame, turning on currents or whatever else is needed for operation. Give the detector 15-60 minutes to stabilize based on previous experience. Using the same oven temperature conditions and data system settings as when the problem occurred, perform a blank run (i.e., no injection) and collect the chromatogram. If excessive noise, baseline drift or other problems are evident, the detector is probably the source of the problem. If the baseline is flat and there is no evidence of the problem, the detector is not the source of the problem. The second part of the test is to install a 1-2 meter piece of clean, deactivated fused silica tubing in the GC using the same method and conditions as for a regular column. Do not use an old piece of column. Establish a carrier gas flow similar to the one used with the regular column. Perform the blank run again and collect the chromatogram. If excessive noise, baseline drift or other problems are evident, the source of the problem is most likely the injector, carrier gas or incoming gas lines. The detector has already been verified to be functioning satisfactorily with the first test. Since the column is not installed, it is not responsible for the problem. If the baseline is completely or fairly flat and there is no evidence of the problem, the injector, carrier gas or incoming gas lines are probably not the source of the problem. The problem is most likely associated with the column. The final test is to install and condition the column as usual. Perform the blank run one more time. If the problem is not evident, the problem is solved. Sometimes, the process of re-installing a column corrects the problem. If the problem returns after installing the column, the problem is with the column or installation technique. The column may be not damaged, but only contaminated or improperly installed. If an injector or related areas were found to be contaminated, it is likely that the column may be contaminated also. If the contaminants in the injector have also contaminated the column, solvent rinsing the column may be warranted.

12.4.2   Condensation Test

The condensation test is used when contamination of the injector, carrier gas or gas lines is suspected. Keep the column at 40°C for at least several hours; 8-24 hours is better. Afterwards, temperature program the column at 10-20°/min starting at 40°C and ending at the column conditioning temperature. Collect the chromatogram. As soon as the column has cooled down and the GC is ready, immediately start another blank run using the same conditions. Collect the chromatogram and compare it to the first one. If the chromatograms are similar, it is unlikely that the injector, carrier gas or gas lines are severely contaminated. If the first chromatogram (the one kept at the low temperature for 8-24 hours) is worse than the second chromatogram, there is a contamination problem with the injector, carrier gas and/or gas lines, or significant septum bleed.

By allowing the column to stay at a low temperature, many of the contaminants are trapped (retained) by the column. The longer time at the low temperature, allows more of the contaminants to be collected. When the column is temperature programmed, the collected contaminants elute from the column. More contaminants accumulate when the column is left for many hours at the low temperature (first chromatogram), thus the blank run looks worse than the column with minimal time at the low temperature (second chromatogram). Similar first and second chromatograms mean there was only a small amount of contaminants collected in the column even after hours at the low temperature. The first chromatogram often looks slightly worse than the second one. It is common for a small amount of system contamination and septum bleed to be present. It is large differences between the two chromatograms that indicates a contamination problem exists.

12.4.3   Check Out Column

A check out column is one that is used to evaluate the performance of the GC. The performance of the check-out column should be thoroughly documented for comparison purposes. Upon removal of the original column and installation of the check out column, the original column is eliminated as a problem cause. If the check out column's performance is normal (as previously verified and documented), the original column or its installation is at fault. If a problem is noticed with the check out column installed, the problem is with the GC (providing the check out column was properly installed and set up).

Ideally, the check out column should be the same size and phase as the regular column. If more than one column is used in the GC, use the column with the most non-polar stationary phase for the check out column. It is not essential that the column be the same as one being used since the GC is being evaluated and not the column. A similar column is still best since its performance, characteristics and behavior are familiar. The check out column should be used exclusively for evaluating system performance and not used for sample analysis. A new column makes a satisfactory check out column; however, its performance in the GC has not been previously documented and adds a bit of uncertainty to the check out procedure.

12.4.4   Column Exchange

A column exchange is another method to determine whether a column is the cause of a problem. Another properly functioning column or GC is needed for this test. Ideally, a column with the same description is available. If an installed column is suspect, replace it with the other column. If the problem returns, the GC or the column installation technique is the problem. If the problem does not return, the first column is probably at fault.

Another approach is to install a suspected problem column in a properly working GC. If the problem returns, the column is probably at fault. If the problem does not appear, the original GC is probably at fault. Sometimes the process of re-installing a column fixes the problem, thus re-install the suspected problem column back in the original GC and evaluate before attempting to repair the GC.

12.4.5   Column Test Samples

Column test samples can often help to pinpoint problems. The column manufacturer's test sample can measure activity, efficiency and retention characteristics. A self made test sample that mimics the actual samples provides information about the performance of the GC for a particular analysis. For either type of test sample, a lot of information about the GC is obtained. Injecting and evaluating a test sample should be one of the first tests performed after all of the obvious and simple inspections and tests have been completed. Chapter 10 outlines the use of the various types of test samples and the interpretation of the test results.

12.4.6   Static Pressure Check

A static pressure check is probably the fastest and most complete method to determine whether a leak is present in the injector or the carrier gas lines. A column head pressure gauge or in-line pressure gauge in the incoming carrier gas line is necessary to perform a static pressure check. To perform the test, the column is removed from the injector and the column inlet is sealed with a leak free cap. Also, replace the septum with a new one. For a split/splitless injector system, set the split line in the closed position (purge off) so that a leak in the vent solenoid can be detected. Cap off the septum purge line to prevent gas from escaping from this route. Pressurize the injector by supplying carrier gas. After the pressure has stabilized, eliminate the flow of carrier gas to the injector by turning the carrier gas off at its source (i.e. at the cylinder or in-line pressure regulator). Do not use the flow controller or pressure regulator on the GC to stop the gas flow. Some controllers are set so that they do not completely stop carrier gas flow. Also, some controllers may be easily damaged if they are tightened excessively in the off position. The head pressure should not drop more than 1-2 psi over a 10-15 minute time span. A drop in the head pressure indicates that there is a leak in the injector or the carrier gas lines. The most common sources of leaks are loose fittings on the injector or the incoming gas lines.

12.5   Common Problems and Their Causes

The following comments are generalizations and simplifications. It is impossible to cover every possible problem and solution. Also, there seem to be exceptions and contradictions to nearly every problem, cause and symptom. The causes for each problem are listed in the order of most to least common or likely. Where appropriate, the previous book section pertaining to the explanation, prevention or solution to a problem is provided in parentheses.

12.5.1   Baseline Drift or Wander

Drift or wander is the upward or downward movement of the chromatogram baseline. It can also be a combination of multiple upward and downward movements within a single chromatogram. Make sure not to confuse normal column bleed with a baseline drift or wander problem.

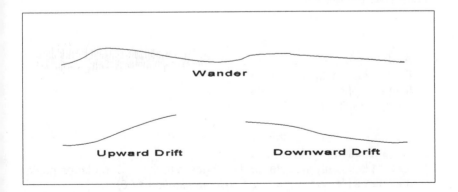

Figure 12-1: Baseline Drift or Wander

1. Contamination - semi-volatile residues eluting from the column. *(Contamination 11.6)*

2. Excessive column bleed - baseline rise should start 30-40°C below the upper isothermal temperature limit of the column. *(Column bleed 4.6)*

3. Change in the carrier gas flow during temperature programming for a concentration dependent or flow sensitive detector. *(Detector sensitivity 8.2.4)*

4. Detectors that have not been allowed to fully equilibrate or condition. This is most common for ECD, TCD and MS systems.

5. Detector difficulties - temperature, current, filament, bead, lamp or gas flow rate irregularities or inconsistencies.

6. Loose, dirty or corroded detector connections.

12.5.2   Noisy Baseline

Baseline noise is always present. It should be small, but visible at sensitive settings. A noise measurement taken during normal and satisfactory operation should be available for comparison purposes.

Figure 12-2: Baseline Noise

1. A defect or expiration of a flame jet, collector, bead, filament, photomultiplier tube, electron multiplier, etc..

2. Column or injector contamination - the noise pattern is usually irregular and changes with time. Sometimes peaks may be visible above the noise. *(Contamination 11.6)*

3. An air leak in ECD, TCD or MS systems especially in the areas around the detector.

4. A column inserted into the flame jet of a FID, NPD or FPD. Usually peak tailing also occurs. *(Column installation in the detector 9.2.6)*

5. Defective detector board or poor connection.

6. Incorrect air and hydrogen ratio for FID, NPD and FPD systems. *(Gases: FID 8.3.2, NPD 8.4.2, FPD 8.7.2)*

7. Outside interferences such as electrical equipment or air currents around the detector.

12.5.3 Spikes in the Baseline
Spikes appear as a single line and not as a triangle (i.e., peak).

Figure 12-3: Spikes in the Baseline

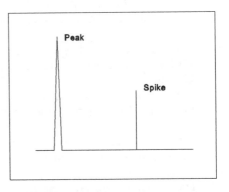

1. Particulate matter passing through the detector. Usually the particles are from a dirty flame detector. Sometimes the particles originate from the stationary phase of PLOT columns.

2. Cycling of other equipment in the same room or electrical circuit. Check the other equipment to determine if the appearance of a spike coincides with a particular event occurring with one of the other pieces of equipment.

3. Loose connections on cables and circuit boards. Thermal expansion and contraction of connections may cause the spiking behavior especially with very old or inexpensive GC's.

12.5.4   Tailing Peaks

Peak tailing is caused by any event that causes some of the sample to travel down the column at a slower rate than the majority of the sample. Also, peak tailing is caused by a portion of the sample entering the column later than the majority of the compound.

Figure 12-4: Tailing Peaks

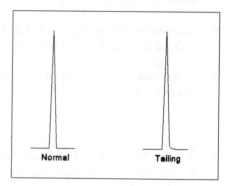

1. Contamination - semi- and non-volatiles residues in the column and injector. Active compounds are more susceptible to tailing problems. More severe cases of contamination also cause less or non-active compounds to tail. *(Contamination   11.6; Tubing activity 3.2; Active compounds 10.2)*

2. Column activity - active compounds are the most susceptible to tailing problems.   Inactive compounds do not tail due to column activity. Usually the tailing is more severe for the later eluting peaks unless the later eluting compounds are significantly less active than the earlier eluting peaks. *(Tubing activity 3.2; Active compounds 10.2)*

3. Backflash - peak tailing is usually more severe for the earlier eluting peaks (more volatile) compounds. *(Backflash 7.2.3)*

4. Dead volumes caused by poorly installed liners, unions or columns. The peak tailing is usually more severe for the earlier eluting peaks (more volatile) compounds.

5. Lack of solvent effect or cold trapping in splitless and on-column injections. *(Splitless 7.4.3 and 7.4.4; On-column 7.7.2)*

6. Too low split ratios. *(Split ratios 7.3.2)*

7. Too low or no detector makeup gas. *(Makeup gas 8.2.2)*

8. Poorly matched stationary phase and sample. Greatest problem when the polarity of the sample compounds and solvent is very different from the column, or when there are large differences in the pH of the sample compounds and stationary phase.

9. Poor injection technique - erratic or discontinuous depression of the syringe plunger. *(Injection techniques 7.8)*

10. Condensation of the vaporized sample on a cooler region in the injector (e.g., septum, incoming gas lines, transfer lines or a poorly heated region of the injector).

11. Plug of glass wool is too tightly packed or too low in the injector liner. *(Packed injector liners 11.9)*

12. The co-elution of two peaks giving the appearance of a single peak that tails.

13. Overloading of a PLOT column *(Column capacity 2.11)*

12.5.5   Fronting Peaks

Fronting peaks are noted by tailing or slanting of the leading edge of the peak. Peak fronting is caused by a portion of the sample entering the column earlier than the majority of the compound. Also, peak fronting is caused by any event that causes some of the sample to travel down the column at a faster rate than the majority of the sample.

Figure 12-5: Fronting Peaks

1. Column position in the injector which is more of a problem with high boiling point compounds. *(Split 7.3.5; Splitless 7.4.7; Megabore direct 7.5.4)*

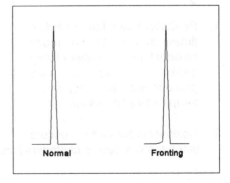

2. Compound and solvent/compound mismatch - occurs when the sample compounds are significantly more soluble in the sample solvent than in the stationary phase.

3. Column overloading - evident as a shark fin shaped peak instead of a curved front peak. *(Column capacity 2.11)*

4. Poor injection technique - sample in the syringe needle. Greater fronting for the earlier eluting (more volatile) peaks. *(Injection techniques 7.8)*

5. No retention gap for on-column injections with secondary cooling. *(On-column retention gap 7.7.4)*

6. Co-eluting peaks - a small peak in the base of the larger peak.

7. Condensation of the sample in an injector that is too cold.

12.5.6   Extremely Broad or Rounded Peaks

Capillary column peaks should be sharp and pointed at the top or apex. Column efficiencies, injector and system conditions all affect peak width, thus they need to be considered before making judgements on any peak broadening situation.

Figure 12-6: Extremely Broad or
Rounded Peaks

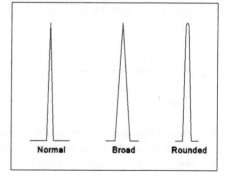

1. Peaks with very long retention times appear to be much broader than the earlier eluting peaks. This is very pronounced for isothermal temperature conditions.

2. Column contamination - occurs when there is severe contamination. *(Contamination 11.6)*

3. No solvent effect or cold trapping for splitless or cool on-column injections. *(Splitless 7.4.3 and 7.4.4; On-column 7.7.2)*

4. No retention gap for on-column injections using secondary cooling. *(On-column retention gap 7.74)*

5. Too low or high carrier gas linear velocity. *(Recommended linear velocities 6.6)*

6. Overloaded detector - a greater problem for detectors with a small linear range. *(Detector linear range 8.2.6)*

7. Co-elution of two compounds appearing as a single broad peak.

12.5.7 Split Peaks

Usually split peaks are caused by sample introduction (injector) problems. A damaged or malfunctioning column can not cause split peaks.

Figure 12-7: Split Peaks

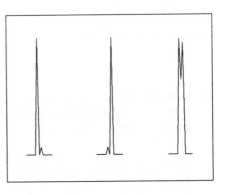

1. Injection technique - jerky or erratic syringe plunger depression. Also, sample in the syringe needle especially when the sample compounds or solvent have low boiling points. *(Injection techniques 7.8)*

2. Column position in the injector - the column is either too low or high in the injector. This is liner and sample dependent. *(Column position in the injector: Split 7.3.5; Splitless 7.4.7; Megabore direct 7.5.4)*

3. Severe backflash - greater problem for the earlier eluting (more volatile) peaks. *(Backflash 7.2.3)*

4. Detector overload - occurs for very large peaks such as the solvent front. The detector becomes "blinded" by the large amount of compound and momentarily decreases in response. The response drop-off creates a small valley which gives the appearance of a split peak.

5. Sample degradation in the injector - usually a separate peak is observed for the degradation product.

6. Column temperature fluctuation during a run - problems with oven heater or a heat source that is not sufficiently shielded from the column. Multiple peaks may be observed. This problem is more common for very inexpensive or old gas chromatographs. If the heating element is exposed, place aluminum foil or a metal shield in front of it to block the direct line to the column.

12.5.8   Negative Peaks

Negative peaks are usually detector or compound related. In some cases, negative peaks are normal.

Figure 12-8: Negative Peaks

1. Compound structure - some compounds generate a negative response with a TCD or ECD.

2. Dirty or old ECD - the negative peak usually immediately follows a large, positive peak. *(ECD negative peaks 8.5.8.4)*

3. Detector overload - a high concentration of the sample "blinds" the detector and the response momentarily and rapidly drops and then returns to normal. This creates a negative peak and is most common for element specific detectors such as ECD, NPD and FPD.

4. Detector quenching - quenching or reduction of detector response by non-responsive compounds. If no responsive compound is present at the same retention time, a negative peak may appear.

5. If all of the peaks are in the negative direction, check for a reversal of the recorder connections or settings.

6. Contamination - contaminants in the TCD reference cell gas lines.

12.5.9 Flat Top Peaks

Some peaks in a chromatogram will have flat tops since they are off scale. Changing the attenuation or scale to get them on scale causes some of the smaller peaks to disappear.

Figure 12-9: Flat Top Peaks

1. Exceeding the maximum input signal of the data system or recorder - increase the range (maximum input signal) of the data system or decrease the detector output signal.

2. Exceeding the maximum output of the detector - increase the range of the detector signal.

3. The attenuation or full scale setting for the data system or recorder is too low - the peaks are consistently off scale and appear to be flat on the top. In reality, the peaks are well formed, but flat-topped due to being off the top of the page or screen.

12.5.10 Peak Size Problems

One to all of the peaks in the chromatogram can change in size. Peaks that are too small are the most common peak size problem. It is important to consider compound structures and properties when evaluating a peak size problem.

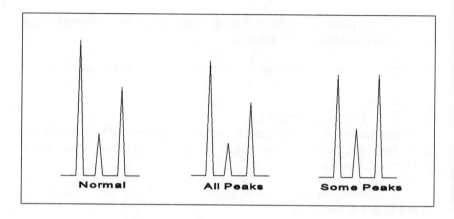

Figure 12-10:   Peak Size Problems

12.5.10.1 No Peaks

1. Detector is not on or properly connected to the recorder or data system.

2. Very low or no carrier gas flow through the column. *(Turning on and verifying the carrier gas flow 9.2.5)*

3. Plugged syringe.

4. If an autosampler is being used, check whether there is enough sample in the vial or whether the autosampler is sampling from the correct position.

5. Injecting into the wrong injector.

6. Broken column. *(Column breakage 11.2)*

7. Column installed in the wrong injector or detector.

12.5.10.2 All Peaks

All of the peaks changed in size; however, the change may not be equal for all of the peaks.

1. Change in the syringe technique - for example, not accounting for the volume of sample contained in the syringe needle or any residual solvent in the syringe from a previous injection or cleaning. *(Injection techniques 7.8)*

2. Change in the split ratio. *(Split ratios 7.3.2)*

3. Change of the purge on time for a splitless injection. *(Purge activation times 7.4.2)*

4. Change in the septum purge flow - greater effect on the earlier eluting (more volatile) peaks. *(Septum purge flows: Split 7.3.3; Splitless 7.4.5; Megabore direct 7.5.3)*

5. Change of detector response - often does not affect all compounds equally. Incorrect type or flow of detector gases, temperatures, settings or detector contamination. *(Detector sensitivity 8.2.4)*

6. Leaks in the injector - greater affect on the earlier eluting (more volatile) peaks. Usually accompanied by peak tailing. *(Leak detection 9.6)*

7. Change in the sample - degradation, reaction or loss of one or more of the sample compounds. Error in sample preparation or extraction.

8. Backflash - loss of sample via the septum purge line; a greater problem with the earlier eluting (more volatile) peaks. Condensation of compounds on a cool spot; a greater problem for late eluting (less volatile) peaks.

9. Partially blocked or clogged syringe.

10. Activity - adsorption of active compounds by the column or injector liner. *(Tubing activity 3.2; Active compounds 10.2)*

11. Contamination - adsorption of active compounds by contaminants in the column or injector liner. *(Contamination 11.6)*

12. Too low of an injector temperature - greater affect on the later eluting (less volatile) peaks. *(Injector temperatures 7.2.1)*

13. The GC was started while the baseline was dropping or drifting downward. Only the tops of the peaks are visible.

14. Incompatible compound - poor compound response with the detector. Some detectors are very sensitive to small changes in compound structure.

12.5.10.3 Some Peaks

1. Activity - adsorption of active compounds by the column or injector liner. *(Tubing activity 3.2; Active compounds 10.2)*

2. Contamination - adsorption of active compounds by contaminants in the column or injector liner. *(Contamination 11.6)*

3. Change in the sample - degradation, reaction or loss of one or more of the sample compounds. Error in sample preparation or extraction. A pH change may create a salt or ionized form of an acid or base that may not chromatograph properly.

4. Change of detector response - often does not affect all compounds equally. Incorrect type or flow of detector gases, temperatures, settings or detector contamination. *(Detector sensitivity 8.2.4)*

5. Backflash - loss of sample via the septum purge line or condensation on a cool spot. Greater problem with the earlier eluting (more volatile) peaks. *(Backflash 7.2.3)*

12.5.11   Retention Time Changes

1. Change in carrier gas linear velocity - the peak retention times change, but their k values do not change. *(Measuring linear velocities 6.4.1)*

2. Carrier gas linear velocity set at a different column temperature. *(Measuring linear velocities 6.4.1)*

3. Change in column temperature - not all compounds are affected by the same amount.

4. Change in the sample solvent.

5. Large change in one or more of the sample compound concentrations.

6. Leak in the septum - usually causes erratic retention times. *(Leak detection 9.6)*

7. Column contamination - interaction of the sample with the contaminating residues. *(Contamination 11.6)*

12.5.12 Loss of Resolution

A decrease in peak separation, peak tailing, or an increase in peak width can all be the actual cause of a loss in resolution. Determine which has occurred before troubleshooting the GC system.

Figure 12-11: Loss of Resolution

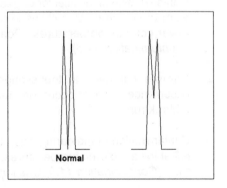

1. Contamination of the column - semi- or non-volatile residues deposit in the column. Usually peak broadening occurs, but peak separation decreases occur in some cases. *(Contamination 11.6)*

2. Damage to the stationary phase - oxygen, thermal and/or chemical induced damage to the stationary phase. Usually, higher column bleed or activity is also experienced. *(Oxygen damage 11.4; Thermal damage 11.3; Chemical damage 11.5)*

3. Change in carrier gas linear velocity - check the peak retention times as an indicator of a shift in the linear velocity. *(Measuring linear velocities 6.4.1)*

4. Large changes in the sample concentration - usually causes a broadening in the peak width. A change in the peak retention time may occur also.

5. Normal column degradation with use.

6. A different column.

7. Change in the column temperature.

12.5.13   Ghost Peaks or Carryover Problems

Columns can not generate peaks without the injection or introduction of compounds. If the ghost peaks are similar in width to the ones in the injected sample, the contaminants probably entered the column with that injection. If the ghost peaks are significantly wider, the contaminants probably entered the column with a previous injection or were carried into the column with the carrier gas.

1. Contamination of the injector or gas lines - semi-volatile residues eluting from the column. Usually becomes worse the longer the GC is left at lower column temperatures. Peaks are usually visible in a blank run. *(Contamination 11.6)*

2. Sample carryover in the syringe - variable ghost peak sizes and occurrence. Should occur with the injection of clean solvent, but not in a blank run.

3. Contamination of the carrier gas - becomes worse the longer the GC is left at lower column temperatures. Peaks are usually visible in a blank run. *(Contamination 11.6; Gas impurity traps 11.10)*

4. Contamination of the sample solvent - ghost peak size and occurrence are dependent on the injection volume. No peaks should be visible in a blank run.

12.5.14 Excessively Broad Solvent Front

Sometimes, a wide solvent front may be normal. Comparisons should be made with previous results before assuming a problem exists.

Figure 12-12:   Excessively Broad
                         Solvent Front

Normal

1. Purge function not used with a splitless injection - extremely wide (>4 minutes) solvent fronts. *(Purge activation time 7.4.1 and 7.4.2)*

2. Too long a purge activation (on) time. *(Purge activation time 7.4.1 and 7.4.2)*

3. Too low a split ratio. *(Split ratios 7.3.2)*

4. Using high boiling point solvents at low initial or isothermal column temperatures.

5. Excessively large injection volumes (greater than 3 µL) especially for direct, on-column and splitless injections.

6. A leak in the injector - a greater problem with low boiling point sample solvents. *(Leak detection 9.6)*

7. Too low an injector temperature - a greater problem with high boiling point solvents. *(Injector temperatures 7.2.1)*

8. A poorly installed column in the injector. *(Column installation in the injector 9.2.4)*

9. Very slow depression of the syringe plunger - a greater problem with low boiling point sample solvents. *(Injection technique 7.8)*

12.5.15   Quantitation Problems

1. Activity - adsorption of active compounds by the column or injector liner. Greater impact on lower level compounds. *(Tubing activity 3.2; Active compounds 10.2)*

2. Contamination - adsorption of active compounds by the column or injector liner.   Greater impact on lower level compounds. *(Contamination 11.6)*

3. Contamination - peak tailing or shape distortion, thus altering peak integration. *(Contamination 11.6)*

4. Contamination - causes baseline disturbances such as noise, wander, drift, high background signal, negative peaks or interferences. Interferes with peak integration. *(Contamination 11.6)*

5. Change in the syringe technique. *(Injection techniques 7.8)*

6. Injector discrimination due to a change in injector conditions. *(Discrimination 7.2.4)*

7. Exceeding the linear range of the detector. *(Detector linear range 8.2.6)*

8. Incorrect detector conditions - detector gas flows, temperatures, voltages or currents.

9. Improper recorder or data system settings - incorrect or inconsistent peak integration.

10. Improper sample handling or storage - changes in the sample concentration from evaporation, reaction or compound degradation.

12.5.16   Rapid Column Deterioration

1. Contamination - deposition of high molecular weight contaminants not removed by solvent rinsing the column. Contaminants polymerized or bound to the column due to high temperature bakeout. *(Contamination 11.6; Solvent rinsing 11.7)*

2. Oxygen damage - exposure of the column to oxygen especially at high temperatures. The most common reasons for the presence of oxygen in the carrier gas are entry of air through a leak or use of poor quality carrier gas. *(Gas purity 6.7; Oxygen damage 11.4; Gas impurity traps 11.10)*

3. Chemical damage - injection of mineral acids and bases. Perfluoro organic acids may also damage stationary phases. *(Chemical damage 11.5)*

4. Thermal damage - exposing the column to temperatures above its upper temperature limit. Longer exposure times, higher temperatures and higher oxygen levels increase the speed and severity of the damage. *(Thermal damage 11.3)*

5. Breakage - excessive abrasion or scratching of the polyimide coating. Breakage is more common for larger diameter columns. *(Column breakage 11.5)*

6. Polar stationary phases have much shorter lifetimes than non-polar phases.

APPENDIX A: TERMS AND ADDITIONAL EQUATIONS

A.1 Terms

α	separation or selectivity
β	phase ratio
CE	coating efficiency (%)
d_f	film thickness
F	flow rate (mL/min)
$\bar{F}$	average flow rate (mL/min)
h	height equivalent to a theoretical plate
H	height equivalent to a effective theoretical plate
I	retention index
K_D	distribution constant
k	partition ratio
L	column length
n	number of theoretical plates
N	number of effective theoretical plates
r	column radius
R	resolution
t_m	retention time of a non-retained compound
t_r	retention time
t_r'	adjusted retention time
TZ	Trennzahl
$\bar{u}$	average linear velocity (cm/sec)
W_b	peak width at half height (in time units)
W_h	peak width at base (in time units)

A.2 Additional Equations

$$W_b = 1.699 \, W_h$$

$$t_r = t_m (k + 1)$$

$$n = 16 \left(\frac{t_r}{W_b} \right)^2$$

$$N = 5.545 \left(\frac{t_r'}{W_h} \right)^2$$

$$N = 16 \left(\frac{t_r'}{W_b} \right)^2$$

$$H = \frac{L}{N}$$

$$R = 2 \left(\frac{t_{r2} - t_{r1}}{W_{b1} + W_{b2}} \right)$$

$$TZ = \frac{R}{1.177} - 1$$

APPENDIX B: TABLE OF PHASE RATIOS

Phase Ratios

Film thickness (µm)	Column Diameter (mm)				
	0.18	0.20	0.25	0.32	0.53
0.10	450	500	625	800	1325
0.15	300	333	417	533	883
0.18	250	278	347	444	736
0.20	225	250	313	400	663
0.25	180	200	250	320	530
0.33	136	152	189	242	402
0.40	113	125	156	200	331
0.50	90	100	125	160	265
0.83		60	75	96	160
1.00		50	63	80	133
1.50			42	53	88
3.00			21	27	44
5.00			13	16	27

APPENDIX C: TEMPERATURE LIMITS

Typical Temperature Limits

Phase	Temperature limit*
Polysiloxanes:	
100% dimethyl	-60 to 325/350°C
100% dimethyl (high temp.)	-60 to 400°C
5% diphenyl-95% dimethyl	-60 to 325/350°C
5% diphenyl-95% dimethyl (high temp.)	-60 to 400°C
20% diphenyl-80%dimethyl	40 to 290/310°C
35% diphenyl-65%dimethyl	40 to 280/300°C
50% diphenyl-50%dimethyl	40 to 280/300°C
50% diphenyl-50%dimethyl (high temp.)	40 to 340/365°C
6% cyanopropylphenyl-94% dimethyl	-20 to 280/300°C
14% cyanopropylphenyl-86% dimethyl	-20 to 280/300°C
50% cyanopropylphenyl-50% dimethyl	45 to 220/240°C
50% cyanopropyl-50% dimethyl	40 to 250/260°C
50% trifluoropropyl-50% dimethyl	45 to 240/260°C
Polyethylene glycols:	
PEG ("WAX")	40 to 250/260°C
Acid treated PEG ("FFAP")	40 to 250/260°C
Base treated PEG ("CAM")	40 to 250/260°C

*Upper limits are usually about 20°C less for d_f > 1 µm.

*Temperature limits may be slightly different for capillary columns from various manufacturers. Actual temperature limits should be confirmed by the specific column manufacturer.

APPENDIX D: COLUMN BLEED MASS SPECTRA

These column bleed mass spectra are representative of ones typically obtained for each listed stationary phase. While the major ions will be the same for most GC/MS systems and similar columns, there will be differences in the identities and abundances of the minor ions. Column manufacturer, GC/MS parameters, column age and history, carrier gas purity, and system contaminants will all affect the minor ions.

The mass spectra were obtained using a Hewlett Packard 5971 GC/MS system and at the isothermal temperature limit of each column.

100% dimethylpolysiloxane

5% diphenyl-95% dimethylpolysiloxane

35% diphenyl-65% dimethylpolysiloxane and 50% diphenyl-50% dimethylpolysiloxane

6% cyanopropylphenyl-94% dimethylpolysiloxane and 14% cyanopropylphenyl-86% dimethylpolysiloxane

50% cyanopropylphenyl-50% dimethylpolysiloxane

50% cyanopropyl-50% dimethylpolysiloxane

50% trifluoropropyl-50% dimethylpolysiloxane

100% polyethylene glycol (PEG)

APPENDIX E: TABLE OF COLUMN CAPACITIES

Approximate Column Capacities (ng per compound)

Film Thickness (µm)	Column Diameter (mm)			
	0.18-0.20	0.25	0.32	0.53
0.10-0.20	20-35	25-50	35-75	50-100
0.25	35-75	50-100	75-125	100-250
0.50	75-150	100-200	125-250	250-500
1.0	150-250	200-300	250-500	500-1000
3.0		400-600	500-800	1000-2000
5.0		1000-1500	1200-2000	2000-3000

Similar stationary phase and compound polarities.

Capacity reached when peak width at half height increased by >10%.

APPENDIX F: CAPILLARY INJECTOR DESIGNS

F.1 Flow and Pressure Controllers

Flow controllers deliver a fixed volume of gas per unit time (e.g., mL/min). As long as there is a sufficient volume of incoming gas, the flow controller delivers the same flow rate of gas. Pressure regulators adjust to maintain the same pressure over a wide range of gas flow rates. If a different amount of gas is delivered to the pressure regulator, it adjusts to keep the pressure the same as before. Needle valves are dependent on the pressure of the incoming gas. At a fixed valve setting, an increase in the pressure results in a greater flow rate through the needle valve; a decrease in the pressure results in a lower flow rate. The flow rate out of a gas line equipped with a needle valve is dependent not only on the setting on the valve, but on the pressure of the incoming gas. At a fixed pressure, the flow rate can be adjusted by opening or closing the needle valve. Opening the needle valve delivers more gas while closing the valve delivers less gas.

F.2 Back Pressure Regulated Injectors

Back pressure regulated injectors are found on most Hewlett Packard and Varian gas chromatographs. Back pressure regulated injectors use flow controllers and pressure regulators to control the various gas flows (Figure F-1). A flow controller on the incoming carrier gas line controls the flow rate of carrier gas entering the injector. This flow rate is equal to the sum of the column, split line and septum purge flow rates. The split line is equipped with a pressure regulator and a pressure gauge that measures the pressure in the injector. When the pressure regulator is adjusted, it restricts or opens the split line. Restricting the split line by adjusting the pressure regulator makes it more difficult for the gas to flow through the split line at the same flow rate.

Figure F-1:   Back Pressure Regulated Injector

This increases the pressure in the injector, thus increasing the carrier gas flow or linear velocity in the column. Opening the split line by adjusting the pressure regulator decreases the pressure in the injector, thus decreasing the carrier gas flow or linear velocity in the column. The pressure regulator functions to control the carrier gas flow or linear velocity in the column. The pressure regulator is usually labelled as the column head pressure or column flow on most GC's since it determines the amount of carrier gas entering the column.

The flow out of the split line is controlled by the flow controller on the incoming carrier gas line. If the flow controller is adjusted to introduce more carrier gas into the injector, the pressure regulator adjusts to release the extra gas and to maintain the same column head pressure. This increases the gas flow through the split line, thus increasing the split ratio while keeping the column flow rate the same. If a lower volume of carrier gas is introduced into the injector, the pressure regulator adjusts to allow less gas to exit the split line and to keep the column head pressure at the same value. This decreases the split ratio while keeping the column flow rate the same. The flow controller on the incoming carrier gas line controls the total flow of carrier gas into the injector, thus it is usually labelled as total flow. Sometimes it is labelled as split flow.

The septum purge line is equipped with a flow controller. The flow out of the septum purge line does not change when the incoming carrier gas flow or pressure is changed. This means that the septum purge flow rate does not change when the column or split flow is changed. Also, the septum purge flow can be changed without affecting the column or split flow.

An advantage of back pressure regulated injectors is the column, split or septum purge flows can be individually adjusted without significantly affecting the others. Large flow changes may cause a small change in one of the other flows, but the changes are usually quite small and of no consequence. Another advantage is the flow to the injector never exceeds the value set with the flow controller on the incoming carrier gas line. If there is a leak in the injector, the total volume of delivered carrier gas remains the same. This is advantageous with hydrogen as a carrier gas. It is nearly impossible to reach the explosive level of hydrogen even with a large leak in the injector or a broken column.

A leak in a back pressure regulated injector may cause split ratios, the amount of injector discrimination or peak sizes to change. The pressure regulator adjusts to maintain the same column head pressure, thus the carrier gas flow or linear velocity does not change. Less gas flows out of the split line since some of it is lost via the leak. This change in the split line flow changes the split ratio, thus the amount of each compound entering the column. Some of the sample is lost with the carrier gas when it escapes from the leak. This affects peak size since less of the sample reaches the column.

During the purge off mode with a splitless injection, the injector purge gas flows across the top of the injector body and out of the septum purge line. Upon sample vaporization and expansion, any sample vapors backflashing to the top of the injector are rapidly carried away by the high flow of gas in this region. Greater losses occur for the more volatile sample compounds, thus altering the amount of discrimination.

F.3   Forward Pressure Regulated Injectors

Forward pressure regulated injectors are found on most Perkin Elmer, Shimadzu and Fisons (Carlo Erba) gas chromatographs. Forward pressure regulated injectors use pressure regulators and needle valves to control the various gas flows (Figure F-2). A pressure regulator on the incoming carrier gas line controls the amount of carrier gas entering the injector. The pressure gauge on the incoming carrier gas line measures the pressure in the injector. As the pressure regulator is adjusted to a higher pressure, the flow or linear velocity of carrier gas in the column increases; decreasing the pressure reduces the flow or linear velocity in the column.

The septum purge and split lines are equipped with needle valves. As the column head pressure changes, the flow through these lines change since the flow through needle valves is pressure dependent. This means that when the carrier gas linear velocity or flow is changed, the split and septum purge flows have to be re-set to maintain the previous flow rates. If the carrier gas flow or linear velocity is increased, the needle valves in the split and septum purge lines have to be closed to lower the flows to the previous rates. If the carrier gas linear velocity or flow is decreased, the needle valves need to be opened further. If the split or septum purge flows are changed, the column carrier gas flow or velocity remains the same. The pressure regulator adjusts to allow more gas into the injector to compensate for the larger amounts flowing out of the septum purge or split lines. This functions to maintain the same pressure in the injector, thus carrier gas velocity or flow in the column.

An advantage of a forward pressure regulated injector is its relative insensitivity to the effects of injector leaks. A leak does not significantly affect the performance of a forward pressure regulated injector. The loss of the carrier gas from the leak is compensated by the pressure regulator. Additional carrier gas is supplied to the injector in an amount equal to the amount lost via the leak. This helps to keep the injector pressure, thus the column flow or linear velocity stays the same. Also, the same septum purge and split flows are maintained. There is loss of vaporized sample from the leak which is the only significant impact on injector performance.

The characteristic that makes a forward pressure regulated injector relatively immune to the effects of a leak is the same characteristic that makes one less than ideal when using hydrogen as a carrier gas. Since the pressure regulator attempts to keep the injector pressure at the same value, any leak causes the amount of carrier gas to be increased until the desired pressure is obtained. If the leak is large or the column is broken near the injector, a very large amount of gas is delivered to the injector and out of the

Figure F-2: Forward Pressure Regulated Injector

leak. The pressure regulator is attempting to pressurize the GC oven and surrounding room to the desired pressure, thus the maximum amount of gas that can be supplied is delivered. This can deliver enough hydrogen to reach explosive levels. Nearly every incident of a hydrogen explosion in a GC oven has occurred with a GC equipped with a forward pressure regulated injector.

During the purge off mode with a splitless injection, the injector purge gas is shut off instead of being diverted as for a back pressure regulated injector. Upon sample vaporization and expansion, any sample vapors backflashing to the top of the injector are carried away by the septum purge gas. This is a much lower flow of gas than for a back pressure regulated injector (split + septum purge). Losses of the more volatile sample compounds upon backflash is much less severe. Still, the greatest losses occur for the more volatile sample compounds.

RECOMMENDED REFERENCES

General Chromatography

1. *C. F. Poole and S. A. Schuette,* "Contemporary Practice of Chromatography", Elsevier, Amsterdam, (1984)

2. *J. M. Miller,* "Chromatography: Concepts and Contrasts", John Wiley and Sons, New York, (1988)

Gas Chromatography

1. *W. Jennings,* "Analytical Gas Chromatography", Academic Press Inc., Orlando, (1987)

2. *L. S. Ettre and J. V. Hinshaw,* "Basic Relationships of Gas Chromatography", Advanstar Communications, Cleveland, (1993)

3. *K. Grob,* "Split and Splitless Injections in Capillary Gas Chromatography, 3rd Edition", Hüthig, Heidelberg, (1993)

4. *K. Grob,* "On-Column Injection in Capillary Gas Chromatography", Hüthig, Heidelberg, (1987)

5. *S. A. Stafford, ed.,* "Electronic Pressure Control in Gas Chromatography", Hewlett Packard Co., Wilmington, (1993)

6. *H. H. Hill and D. G. McMinn,* "Detectors for Capillary Chromatography", John Wiley and Sons, New York, (1992)

INDEX